*Frontispiece:
Author (right) pre-
pares for his first
flight in a Maurice
Farman biplane at
Buc aerodrome, near
Paris, July 2, 1912.
Pilot is Lieutenant
Arthur Noé of the
French army.*

Revised Edition

CONTACT!

The Story of the Early Birds

HENRY SERRANO VILLARD

SMITHSONIAN INSTITUTION PRESS

WASHINGTON, D.C.   LONDON

Library of Congress Cataloging in Publication Data:
Villard, Henry Serrano, 1900–
    Contact!: the story of the early birds.
    Bibliography: p.
    Includes index.
    1. Airplanes—History. I. Title.
TL670.3.V55 1987   629.13'09   87–600170
ISBN 0–87474–947–6

British Library Cataloguing in Publication Data is available

Acknowledgment is made to the following for permission to use copyrighted material:
McGraw-Hill Book Company: *The Papers of Wilbur and Orville Wright*, by Ross A. McFarland; copyright 1953

∞The paper in this book meets the guidelines for permanence and durability of the Committee on Production Guidelines for Book Longevity of the Council on Library Resources.

To Those Who Showed the Way

Foreword

As I have been concerned for a substantial part of the past thirty years with aeronautical publishing, both technical and general, many books and manuscripts on the history of human flight have crossed my desk. I have even dabbled a bit in aviation history on my own account. Being in an age bracket such that my own recollections span the years from 1905 to 1915, I also have acquired some personal knowledge of many of the events that took place between the Wright brothers' first successful flights and the outbreak of World War I. Having thus qualified myself, I can say without reservation that Henry Villard's account of the "early birds" of heavier-than-air flight is one of the most exciting that I have ever encountered.

I read much of the manuscript (after an excellent meal, complete with champagne) while hurtling at some 600 m/hr at a height of 25,000 feet over the Eastern Seaboard. Even to suggest a contrast would be to commit a cliché. But under the circumstances, one could not help making some comparisons. That all this should have happened within my own memory is to me a constant source of amazement.

Although I knew that the author had spent much time in research during the past several years, I was yet astonished by the vast amount of detail that comes to light in this book. The many inventors, pilots, and personalities who appear in his pages emerge as real people, with real human characteristics—not all of them entirely commendable. But, to a man, these pioneers are passionately in love with the idea of flight. In those early days flying was much more an art than it was a science. To many, in fact, it was sheer poetry.

Thanks to the author's long experience in the foreign service, he is able to place his characters not only in proper historical perspective but in settings consistent with the times in which they accomplished their deeds. His story of those magnificent men and women, and their contributions to aviation, will find a permanent place in the record of achievement of human flight.

For readers under thirty, this book will paint a lively, informative, and amusing picture of a time (not really too long ago) when flying was an end in itself—not simply a convenient means of spending a weekend in Paris, Calcutta, or Tokyo. And for those of us who can recall the actual events, it re-creates the sense of excitement we all felt when we first became aware that man was no longer earthbound—that his horizons were, indeed, unlimited.

S. Paul Johnston
Director
National Air and Space Museum
Smithsonian Institution

Preface

The story of the first flying machines and the men who flew them is without doubt one of the most remarkable of the extraordinary century we live in. Those who pioneered the skyways, literally from the ground up, wrote a chapter of daring and discovery often forgotten by the modern traveler who takes the jet age for granted. To bring together the threads of the heroic early days of aviation as they were spun in different parts of the world and, while there is yet time, to weave them into a coherent tapestry of that colorful period is the object of the present volume.

The tale told here is necessarily selective. To record the name of each and every "early bird"; to list the many models of machines that appeared and then disappeared; to enumerate all the successes and the failures, all the achievements of altitude, distance, speed, duration, and rate of climb, would require much more time and space than the average reader would tolerate. Consequently, if some person, place, or event that deserves mention appears to be missing from the context, or if the account seems otherwise incomplete, apologies are in order. But it is believed that most of the highlights, as illustrative of unfolding developments, have been included.

The author is indebted to the Smithsonian Institution, National Air and Space Museum, Mr. S. Paul Johnston, Director, for unfailing courtesy and cooperation in this project; and especially to Mr. Paul Edw. Garber, Senior Curator and Historian, and his able and dedicated assistants. Sincere thanks are extended to them, as well as to Mr. Royal D. Frey, Chief, Research Division, Air Force Museum, Wright-Patterson Air Force Base; to the National Geographic Society; to the Library of Congress, together with the public libraries of the District of Columbia; to Mr. Charles H. Gibbs-Smith, author of various handbooks issued by the London Science Museum; to the Central Museum of Aviation and Cosmonautics, Moscow; and to representatives of the Belgian, Swedish, and Norwegian embassies in Washington who made certain additional contributions possible. For the rest, the author has drawn heavily on his personal recollections and on the memorabilia of the early days of flying in his own treasured collection.

H. S. V.

Introduction

As the world advances into the space age, it is easy to lose sight of the first faltering efforts of mankind to rise above the earth's surface, some seventy-five or eighty years ago. We are so accustomed today to reading about the exploits of astronauts, about orbiting satellites and probes into the cosmos that some of us are even beginning to forget just when the giant leap occurred that landed man on the moon and permanently altered our way of thinking about the universe. We accept the idea of travel through space as readily as we did travel by the early railways, the motor car, the fast transatlantic liners, and, of course, the commercial jet and supersonic airplane.

But the future is built upon the past, and the flying days of the past are one of the imperishable building blocks of history. It is well to remind ourselves of that exciting decade before the First World War when a coterie of gallant men and women braved unknown perils in learning to navigate the air and perfect the art of aviation—forerunner of the voyage in space.

The names of those intrepid pioneers and the machines they flew flashed across the skies of the newborn century like shooting stars, and before becoming lost in the mists of time they deserve to be recalled to the modern generation. That was the main reason why CONTACT! was written in the first place and why it is being reissued after a lapse of nearly twenty years! During that interval progress in the exploration of space has been accelerating faster than science fiction, so that the saga of those who first found a way of getting off the ground and ascending into the heavens takes on an evermore profound significance in retrospect. That—and a proven interest in the subject—argued strongly for the current reprint.

The story of the early birds here recorded makes a fascinating contrast to the latest developments in space technology; it is unique in its wealth of detail concerning the events that tested an aviator's prowess and the records that were set and broken, only to be set and broken again. To this edition has been added a list (see Appendices), not found elsewhere, of the first 100 pilots to be licensed in the principal countries of the world, thus rounding out the picture of an era when anyone who qualified for a brevet was looked up to as a flyer endowed with exceptional ability, a true hero of the air.

Contents

Illustrations

CONTACT!
THE STORY OF
THE EARLY BIRDS

CHAPTER ONE

The Early Birds

NORFOLK, VA., DEC. 18—It is reported here that a successful trial of a flying machine was made yesterday near Kitty Hawk, N.C., by Wilbur and Orville Wright of Dayton, Ohio. It is stated that the machine flew for three miles in the face of wind blowing at the registered velocity of 21 miles an hour, and then gracefully descended to earth at the spot selected by the man in the navigator's car as a suitable landing spot. The machine has no balloon attachment, but gets its force from propellers worked by a small engine. . . .

In these matter-of-fact words halfway down the front page under the heading "Soared Like An Eagle," the *Washington Post* on Monday, December 19, 1903, informed its readers of an event that was to alter the destiny of mankind. Six brief paragraphs were enough—so deep was the journal's skepticism—to report that on Saturday, December 17, the problem of mechanical flight had to all intents and purposes been solved. The lead story was the centennial celebration of the Louisiana Purchase. France's Ambassador Jusserand and David R. Francis, president of the St. Louis Fair, were among those taking part in the ceremonies at New Orleans, which included a "Gorgeous Colonial Ball" and a review of the fleet of visiting French and American warships.

The rest of the news also dwarfed in interest the quaint item listed in the index on page 1 as "An Airship That Flies." John A. Benson, a wealthy San Francisco real estate operator, had been arrested for bribery by a Secret Service agent at the Willard Hotel and charged with paying $500, for information, to a division chief in the General Land Office. Sudden wealth had accrued to the Vatican—the late Pope Leo was found to have cached $1,850,000 in a hole in the wall of his apartment, to be turned over to the new Pope four months after Leo's death. The Dixie Flyer of the Illinois Central, bound for New Orleans, was wrecked and burned shortly after leaving St. Louis.

Of the many mechanical marvels that ushered in the twentieth century, none was to gratify human aspiration more than the aeroplane. For ages man's dream of imitating the birds, of soaring into space and winging his way to distant parts, had gone unrealized. It had seemed impossible to rise above the earth or to conquer the laws of gravity without help from a bag of hot air or gas. Then, overnight, the era of power-driven air-craft—as distinct from balloons that drifted like gossamer at the will of the wind—had become a fact. The concept of flight had been revolution-ized by the advent of a practical, lightweight gasoline engine.

In that colorful Edwardian decade, from the day the Wright brothers made their first flight to the outbreak of war in Europe in 1914, man learned how to fly. Without benefit of textbook or teacher, he learned by trial and error. He learned in strange contraptions that bore no resem-blance to any craft seen before or since; he learned in monoplanes, bi-planes, and triplanes of various shapes and sizes. He powered his frail machines with motors of 30 or 40 hp and of uncertain performance. He rolled along the ground, or practiced "grass cutting," or made hops of ten, twenty, or a hundred feet. As he mastered the controls, he learned to skim the treetops, cut figure eights, and climb precariously to the clouds —there to switch off the engine and descend in a volplane to what he hoped would be a graceful landing. Again and again he risked his neck to win the plaudits of the crowds below. In rapid succession he set new marks for altitude, distance, duration—and bones were broken as often as the records.

Those were the kindergarten days of aviation, a time of elementary education. Flying was a sport, an exhibition of nerve, a hobby for a mil-lionaire. As such it represented a glamorous, romantic adventure—a lure to fame and, occasionally, fortune. The first fluttering birdmen were renowned for their daring and perseverance in the face of failure. Yet their approach was scientific as well; for like navigators on an uncharted sea, they sought to probe the newly opened wilderness of the skies—using a wholly untried method of locomotion. To explore the ways and experi-ment with the means took the best of their minds. Step by step the pio-neers (some of them formerly bicycle riders and motorcar racers) gained confidence, often becoming practiced with the aid of kites and gliders. With astonishing quickness knowledge increased, experience and skills were acquired. And if the flying machine was primarily a vehicle of science and of sport, its military potential, too, was not wholly overlooked.

In that improbable era aeroplanes were made of sticks of wood, strips of treated fabric, and piano wire. Assembly lines were unthought of; con-struction was a do-it-yourself job—with or without the help of friends. Once in the air, pilots crouched low in open-cockpit monoplanes or sat bolt upright between the struts of biplanes, fully exposed to the elements. Most aviators knew better than to fly in windy weather; a simple test was to see if the smoke from a cigarette went straight up. Progress was spurred by wealthy patrons of the pastime and by publishers of news-papers conscious of the circulation value in races or contests. Tourna-ments were arranged and feats of prowess advertised, while enthusiastic throngs took to the grandstands to watch.

Individually the pioneers were as diverse in background, temperament, and personal characteristics as life itself; collectively they were something like a fraternal order, bound together by the mysteries of flight, by a sense of fatalism, by the hero worship they received from an earthbound public. While the Anglo-Saxon aviator was usually clean shaven, his continental European counterpart rejoiced in beard and waxed moustache. In France particularly, some of the flyers wore long, pointed moustaches which, to prevent them from flapping in the propeller blast and interfering with clear vision, actually had to be held down in a kind of elastic harness. When the Wrights first flew, the garb of an aviator was nothing more nor less than a business suit—sometimes distinguished by golf knickers or the stiff, high collar of the time. But lack of a windshield made goggles a necessity, and to offset the cold breezes a muffler was in order. For the more casual, a cap with the visor turned rakishly to the rear was a badge of the profession. Silk scarves and turtleneck sweaters also were in vogue. Soon helmets and ear flaps, gauntlets and leather jackets, were in style; seasonal protection from the weather became essential. Clothes might not make the man—nevertheless, enterprising merchants began to promote the raiment considered most suitable for successful flight.

Not the least of the disappointments with which the early birds had to contend were the disparaging remarks of those who refused to believe—even in the face of successful demonstrations—that human flight was possible. "You can no more do that than you can fly" was an aphorism to which, it was supposed, there was no possible reply.

In January 1906, more than two years after the Wrights had flown at Kitty Hawk, Alliott Verdon Roe, who became one of England's greatest pioneers, wrote to the London *Times*. Referring to the work of the Wrights, he described his own pet projects and urged British initiative in the field of flying. The *Times* published the letter but added in a footnote: ". . . it is not to be supposed that we in any way adopt the writer's estimate of his undertaking, being of the opinion, indeed, that all attempts at artificial aviation on the basis he describes are not only dangerous to human life, but foredoomed to failure from the engineering standpoint."

That same year the scientist Simon Newcomb declared: "The demonstration that no combination of known substances, known form of machinery, and known forms of force can be united in a practicable machine by which men shall fly seems to the writer as complete as it is possible for the demonstration of any physical fact to be."

Not every man of science, however, was so shortsighted. William Thomson of Scotland, who was knighted as Lord Kelvin for his contributions to physics, mathematics, and electricity, prophesied to a friend some years before his death in 1907: "Young man, if you live to be as old as I am, you will be able to breakfast in New York and dine in London. Nothing can now prevent the development of the aeroplane in a few years." Thomas A. Edison, studying the lifting power of propellers in 1903, observed in more practical vein that "the flying machine is bound to come, but it will take some time at the rate we are going."

Santos-Dumont Model XIV bis

Curtiss biplane

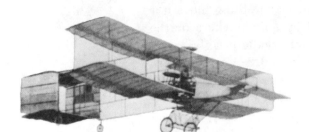

Henry Farman biplane

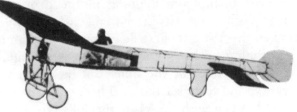

Blériot one-seater military monoplane

Wilbur Wright two-seater biplane

Antoinette monoplane

Grade monoplane

Henry Farman biplane

Maurice Farman biplane

Maurice Farman biplane

Vinet monoplane

Bréguet biplane

Rumpler Taube monoplane

Astra triplane

Wright Astra biplane

Morane-Saulnier monoplane

Caudron biplane

Deperdussin monoplane

Esnault-Pelterie monoplane

Nieuport monoplane

Yet in 1910, when people could no longer deny that the aeroplane had arrived, the celebrated American astronomer William H. Pickering lectured about the possibilities of long-distance flight:

The popular mind often pictures gigantic flying machines speeding across the Atlantic and carrying innumerable passengers. It seems safe to say that such ideas are wholly visionary. . . . Another popular fallacy is to expect enormous speed. Since the resistance of the air increases as the square of the speed and the work as the cube it is clear there is no hope of competing for racing speed with either our locomotives or our automobiles.

Even the name of the new device was a matter of controversy. At first, anything that flew was loosely called an *airship*. Next, for want of a better name, the term *flying machine* was applied to all craft that were heavier than air and that lifted and sustained themselves in the air by mechanical means. Soon the word *aeroplane* appeared—derived from the Greek words *aer* and *planes,* meaning "air mover" or "that which moves in the air." Throughout the early years of flight the use of *aeroplane* was universal. Then, in 1911, it was questioned. A Cambridge philologist, William Walter Skeat, in a letter addressed to the British publication *Flight,* pointed out that *aeroplane* was an awkward compound and suggested that the machine be called an *airplane*. "The opinion of this learned professor of Anglo-Saxon always commands respect and attention," replied the journal, "but we fear 'aeroplane' is too firmly established now to be ousted." It was a good many years before the professor's idea received sympathetic consideration; but gradually the simpler word *airplane* came into use, and is now the official designation in the United States. *Aeroplane* is still widely heard in Great Britain.

In the early 1900's only the dedicated few took the aeroplane seriously, to the extent of predicting a useful future for it as a transport for passengers, mail, or cargo. Most people believed that lighter-than-air craft were the only practical means of traversing the skyways— the dirigible, rigid or semirigid, for regular travel; the free balloon for adventure. The courageous handful who supported the idea of a powered machine heavier than air or who sponsored international competitions for its development were regarded as crackpots and eccentrics.

But great social and economic changes were beginning to be projected on the screen of events to come. Whether generally noticed or not, science—particularly physics—was attaining a stride that would soon alter the world's habits of thinking. The Wright brothers were part of the new era, along with Edison and the electric light, Marconi and his wireless telegraph, De Forest and the audion tube, and Durant, Ford, and many others with their automobiles—not to speak of the early entrepreneurs of the telegraph, the telephone, and the motion picture. Horses, carriages, coachmen, and grooms might still typify the means of getting about; but it would not be long before the motorcar, the motorbus, and the motorcycle would supplant them. The flying machine was close behind.

As the famed Russian flyer and constructor Igor Sikorsky has observed in his "Recollections and Thoughts of a Pioneer," Paris at that time was considered the center of the aviation world:

Aeronautics was neither an industry nor even a science: both were yet to come. It was an "art" and, I might say, a "passion." Indeed, at that time it was a miracle. It meant the realization of legends and dreams that had existed for millenniums and that had been repeatedly pronounced by scientific authorities to be impossible. Therefore even the brief and unsteady flights of that period were deeply impressive. Many times I observed expressions of exultation and tears in the eyes of witnesses who, for the first time, watched a flying machine carry a man in the air.

Like the average motorist, who seldom knows what goes on under the hood, not all the men and women who took up flying were natural mechanics. Sportsmen and sculptors, artists and lawyers, architects and businessmen, could not as a rule be expected to understand the intricacies of internal combustion—or to know how to make repairs when necessary. Many of them learned the hard way, in the school of rugged experience; but many more were heavily dependent on their mechanics. The latter, in turn, were not necessarily experts on the types of motors newly developed for flying machines—yet everything could depend on their ability or intuition in detecting trouble and making repairs. Justly or unjustly, the mechanic was often blamed for a mishap when the pilot himself would scarcely have known how to change a spark plug or clean the valves of his motor.

In any case the mechanic breed was indispensable. To start the motor and apply castor oil—the best lubricant—to the bearings, help was essential. In a day when the hand crank was the sole means of getting an automobile going, an aeroplane could be given life only by turning over the

Before flying was taken seriously, it was pursued as a stunt, for sport, and in a spirit of adventure. Here an early biplane sails through the air at Issy-les-Moulineaux, France, in 1909.

propeller. In the absence of brakes, wooden chocks were placed in front of the wheels; or, in the absence of chocks, a squad of volunteers would hang on to wings, fuselage, and tail while someone turned the propeller slowly by hand to suck fuel into the cylinders. Then—contact! With ignition on and throttle open, the propeller would be swung again. Perhaps there would be an answering roar and a blast of blue smoke; but more often than not, the ritual would have to be repeated. When the motor took hold, the swinger would dart nimbly out of the way and the aeroplane would turn into a vibrant, quivering thing, straining to soar as soon as its attendants released their hold.

There was something else in the personal makeup of the first flyers that set them apart from their fellow men: a fascination with flight that fused a sense of escape with a sense of fulfillment. This new sense of self-expression, of inward exultation, was an emotional experience imparted only to those who could literally rise above the mundane world—who could revel in the freedom of the upper reaches and, from their superior vantage point, gaze down upon the creatures here below. This feeling, which was akin to a fantasy realized, gave rise to an element of chivalry in the early days of World War I, when flyers met in combat to test one another's skill and resourcefulness.

When Henry Farman made his first flights, he told his family, "Your chief sensation is a sort of indescribably joyous feeling that you are independent of the earth." The poetic implications of that sensation were best conveyed in later years by France's great literary prize winner and pilot Antoine de Saint-Exupéry, who was lost on a mission in World War II. Gabriele d'Annunzio, Italy's passionate poet and author, rhapsodized over his first experience as an aeroplane passenger in 1909: "Until now I have never really lived! Life on earth is a creeping, crawling business. It is in the air that one feels the glory of being a man and of conquering the elements. There is the exquisite smoothness of motion and the joy of gliding through space—It is wonderful!"

The glory of being a woman and of conquering the elements was expressed a bit differently by the English writer Gertrude Bacon after she made a flight with Roger Sommer the same year:

The ground was very rough and hard, and as we tore along, at an increasing pace that was very soon greater than any motor I had yet been in, I expected to be jerked and jolted. But the motion was wonderfully smooth—smoother yet—and then—! Suddenly there had come into it a new, indescribable quality—a life—a lightness—a life! Very many there are now who know that feeling: that glorious, gliding sense that the sea-bird has known this million years, and which man so long and so vainly envied it, and which, even now, familiarity can never rob of its charm. . . . You wonderful aerial record breakers of today and of the years to come, whose exploits I may only marvel at and envy, I have experienced something that can never be yours and can never be taken away from me—the rapture, the glory and the glamour of "the very beginning."

The wonders of a new era were illustrated by this intersection near Dayton, Ohio, showing eight forms of transportation. (Popular Mechanics)

Such sentiments are hardly associated with a routine trip on a jetliner today, but in most cases they were felt by pilot and passenger alike in the first hectic days of free flight. What they signified was a new perspective for man—one that would change profoundly his view of the world. Nor were the lyrics confined to passengers on their first ascent. Expounding upon the joys of flight, the 1911 advertisement for one school proclaimed:

To become a companion of the birds; to search the skies and from great heights to look down upon the flattened earth while his monoplane bears him where his whim directs; to realize, to the throbbing of the motor and the song of the propeller, the dream of men throughout the centuries; all of these and more are what flying means. And there is none, except the mentally or the physically unfit, who may not taste its delights.

This sales talk was designed to catch the spirit of the sport—and it did, to a certain extent. But there were deterrents, too. For one, the list of fatalities lengthened remorselessly. Experience was lacking; and where the spirit was willing, the machine was often weak. As against the comparative security of lighter-than-air "vehicles," as one authority put it, the hazards attending the use of heavier-than-air craft were all too obvious. For the most adventurous, "gliding might be admissible as a sporting novelty"; but flight in motor-driven aeroplanes, it was predicted, was "doomed by its insurmountable technical limitations to be no more than a passing fad."

The price paid for each new triumph was indubitably high. Unfamiliarity with conditions aloft and breakage of some vital part in flight were two frequent causes of disaster. Wind, weather, and fragility: the roll of aviators who met a premature end owing to these hazards was a noble as well as a notable one. The fact remains that accidents never held back the development of aviation. Pilots knew the risks and accepted them as part of the game.

Although the earliest attempts to fly were inspired by a simple desire to solve the problem of mechanical flight, they were soon stimulated by the prospect of prizes of silver cups or cash. The lure of a coveted trophy, often elaborately designed and inscribed, did much to encourage competition. Not only was it good public relations for aspiring aviators to enter national or international contests, but the resulting rivalry was sure to improve the know-how of both pilot and constructor. To draw the crowds, a program would be offered that brought together the most noted birdmen of the day. They vied with seeming recklessness for the honors, and for days at a time spectators lived on thrills, chills, and spills. Eyes grew moist witnessing the proof that men could fly; hearts skipped a beat with each miss of the sputtering motors; voices trembled with excitement discussing the spine-tingling crackups and the smashing of world speed or altitude marks.

The famous first meets lent luster to the names of places where they were held—Los Angeles, Boston, Belmont Park in America; Rheims, Doncaster, Blackpool, Lanark, Bournemouth abroad. As cross-country flying became commonplace, point-to-point races were held: London–Manchester, Paris–Madrid, Paris–Rome, the Circuit of Europe, the Circuit of Britain, the Circuit of Germany, all came to be celebrated as competitions. Not only the winners but the also-rans, by participating in the struggle for first place and profiting from the lessons they learned, helped to advance the art of flying and the science of aeronautical engineering.

Useful as they might have been, unofficial or official subsidies, such as airmail contracts, were not forthcoming until much later. But the military in most countries foresaw the possibilities in using aircraft for scouting purposes. Trials tailored to government specifications were held at intervals, and the hoped-for orders connected with them were a prod to innovation and improvement. Out of the many experimental models were developed the combat and bombing planes of World War I—and from these are descended the commercial aircraft of today.

During that prolific first decade, public interest in flying—like that in the Vanderbilt auto races and their Gordon Bennett counterparts—rose to high pitch. As faith in the future of the aeroplane grew, it became fashionable to attend the races and competitions, and even to seek a flight as passenger. To have been "taken up" for a spin was the ultimate status symbol of the day. The potential peril cast the passenger in a role almost as heroic as the pilot's; and the many who returned safe and sound from cloudland not only awed their friends but spread the news of a unique pastime. It was not long before women followed their men into the passenger seat, and from there learned to manipulate the controls. In the earliest schools of flying were female pupils; and some of them became as proficient and well known as the most expert of their brothers.

It was not only on land that the early birds found a roost. In the United States some of the first experiments took place over water; in France, England, Switzerland, and elsewhere the aeroplane equipped with pontoons instead of wheels—the hydroaeroplane—attracted its share of enthusiasts. The prototype of the "flying boat," or aircraft with a hull, which gradually developed into transports such as the Pan American Clipper, made its appearance at about the same time. While these machines never possessed the maneuverability or versatility of the land-based models, they added to the knowledge of aerodynamics and served a purpose of their own in expanding the sport of flying.

Although a number of inventors who followed the Wrights managed to get their ungainly offspring into the air (it was said that a piano could fly if equipped with wings and motor), only a handful could claim equal success. Many models remained grounded or capable only of tentative hops —fated to be the sideshow freaks in the main spectacle of advancing technology. Nevertheless they too made a contribution: the unfettered experiments that took place in the dawn of flight led to the discovery of important basic principles, to be incorporated in later design.

The "good old days," when to fly was merely to find a new thrill in living, are—like the era of the steam locomotive and the horse-drawn fire engine—part of the nostalgic past. As Winston Churchill was to say about the relative peace and security of that prosperous period: "The old world in its sunset was fair to see." But the accelerator of war was abruptly applied to technological development. At the close of Europe's "Belle Epoque," rung down by the guns of midsummer 1914, flying for sport or for record taking was eclipsed by employment of the aeroplane in combat. France and Germany were ready with squadrons of machines that were put to immediate military use, and some hair-raising exploits were features of the first battles. In the United States, for the time being, stunt flying and barnstorming continued; not until America herself became a belligerent did the spotlight shift to military use of the aeroplane. But the old order had ended. Under the imperatives of a world at war, new techniques were applied and production was vastly increased. No longer was the flying machine a crude toy, an experimental apparatus for the aerial engineer; out of the stresses and strains of the war years it was to emerge a fully accepted mode of transport. Its general form may not have altered significantly during that time. But when peace came the novelty had worn off, and vistas were opened to flying such as had been dreamed only by the poets. "Saw the heavens fill with commerce," as Tennyson said in his "Locksley Hall" as far back as 1842, "Argosies of magic sails,/ Pilots of the purple twilight, gliding down with costly bales."

In the next fifty years the commercial airliner was to become part of our way of living; the bomber plane was to rain terror from the skies in a still greater world cataclysm as well as numerous limited wars; and the exploration of space with rockets and capsules was to expand at a dizzying rate. But an imperishable record remains—of success and failure, of dogged effort and spectacular reward, of astonishing progress in tune with the remarkable twentieth century. It is the story of the early birdmen and their flying machines.

CHAPTER TWO

The Spreading of Wings

When the *Washington Post* reported that the Wright flying machine had "soared like an eagle" at Kitty Hawk, the story was not accurate in all respects. There were two six-bladed propellers, according to the *Post,* "one arranged just below the center of the frame and so gauged as to exert an upward force when in motion and the other extended horizontally to the rear from the center of the car, furnishing the forward impetus." There were indeed two propellers (of two blades each, not six)—but they worked from the rear of the main plane instead of in the imaginary positions mentioned. "The machine gradually rose until it obtained an altitude of sixty feet," said the *Post.* In fact it pursued an erratic course never more than ten feet above the ground. Instead of "gracefully" landing on "the navigator's" chosen spot, the twin-propellered biplane came down because of gusty winds and the inexperience of the operator. Instead of "a navigator's car," Orville Wright operated from a prone position next to the motor on the lower wing. And instead of 3 miles, the distance covered was 120 feet, in a time of 12 seconds—"a flight very modest compared with that of birds," in the words of the pilot, but "nevertheless the first in history in which a machine carrying a man had raised itself by its own power and sailed forward on a level course without reduction in speed." The second, third, and fourth attempts that day were not quite so modest; the last one, with Wilbur at the controls, covered 852 feet and lasted 59 seconds.

Only a handful of newspapers bothered to mention the incident at all. Its implication for the future was unperceived, its impact on the public minimal. The Norfolk *Virginian-Pilot,* which heard the story from a telegraph operator who sent a message from the Wrights to their father in Dayton, was alone in giving the report any prominence: a banner headline across page 1 of its December 18 issue. Offered to twenty-one other papers by the *Virginian-Pilot*'s correspondent, the story was bought by only five; and of these only two—the *New York American* and the *Cincinnati Enquirer*—saw fit to publish (and then inaccurately) the brothers' achievement on the day following its occurrence. Even Dayton's *Morning Journal* failed to note that the home-town boys had made good, while the afternoon *News* merely mentioned the episode along with local gossip in

The world's first successful aeroplane, designed and operated by Wilbur and Orville Wright, was a simple biplane construction with twin propellers mounted at the rear. (Air Force Museum)

an inside column. The Associated Press thought the whole business worth less than 350 words.

The astounding thing about the Wright Flyer was not that it had flown but that so little attention was paid to it. The main reason of course was that very few people believed human flight was possible; the unlikely tale served up from the windswept hamlet of Kitty Hawk could not be swallowed whole. And hardly anyone, the inventors included, could foresee the importance the airplane would have for the world. Nobody had the wit to answer the scoffers with Benjamin Franklin's quip, delivered in 1783 when doubt was expressed by a French army officer as to the usefulness of the first balloon: "Of what use is a new-born baby?"

There were, to be sure, witnesses—John T. Daniels, W. S. Dough, and A. D. Etheridge of the nearby Kill Devil Life Saving Station; W. C. Brinckley of the town of Manteo; and a boy from Nags Head named Johnny Moore. For permanent proof Daniels snapped a priceless photo with Orville's camera, arranged beforehand on a tripod. While the happenstance of their presence at Kitty Hawk on December 17 was to make them marked men for as long as they lived, no one at the time took the trouble to run them down and interview them to separate truth from rumor. America's famed journalistic enterprise was, indeed, conspicuously missing: weeks passed before wary newsmen cared—or dared—to print the details of the first flight. Months after publication the facts languished in a limbo of disbelief. And for years to come, the idea that man could fly would be the stock-in-trade of jokesters who thought they knew better. "Darius Green and His Flying Machine," an earthy poem written by John Townsend Trowbridge in 1880, had been incorporated in McGuffey's *Reader* and had become part of the American literary gospel, to ridicule anyone "who believed he could fly and tried to do it."

The newspaper editors, however, had good reason to be cautious. Repeatedly, similar claims had been made and proved false. Only nine days before, on December 8, the impracticability of a man-carrying powered aircraft had been demonstrated by no less an authority than Professor Samuel Pierpont Langley, the secretary of the Smithsonian Institution. His gasoline engine driven "aerodrome" had tumbled ignominiously into the Potomac from its launching platform and nearly drowned the pilot, Charles Matthews Manly. "Airship Fails To Fly," observed the unsurprised *Washington Evening Star*. "Prof. Langley's Machine Goes To River Bottom." With such demonstrable "proof" at hand, it was safer to withhold comment than to risk ridicule in print.

The public disbelief in flight, however, extended only to heavier-than-air craft. Dirigible balloons, filled with hydrogen gas and fitted with engine, propeller, elevator, and rudder, were known to have operated successfully. Captain Thomas Scott Baldwin in the United States, Alberto Santos-Dumont in France, Count Ferdinand von Zeppelin in Germany, were building and flying airships without great fanfare or furor. To the uninitiated, one aircraft was much the same as another. Doubtless it was disconcerting to the Wrights to have their invention dismissed as just another "airship." Yet it would be a long time before the Kitty Hawk Flyer—today the prime exhibit of the Smithsonian's National Air and Space Museum in Washington—would be acclaimed as having truly "opened the era of aviation," and even longer before this national treasure found its final resting place. Because of a controversy with the Smithsonian Institution (which had slighted the Wrights in labeling Langley's reconstructed machine "the first man-carrying aeroplane in the history of the world capable of sustained free flight"), Orville Wright in 1928 sent the original Flyer on loan to the Science Museum in South Kensington, London. Not until 1948, after handsome public amends by the Smithsonian, was it finally brought back to this country.

The primitive biplane, with flexible front rudder (or "elevator") and twin rudders in the rear, that started down a wooden monorail track under its own power on that frosty December morning was lifted into the air by a small gasoline motor built at the Wright Cycle Shop in Dayton. The two propellers, about ten feet apart and connected to chains running over sprockets as in a bicycle, revolved in opposite directions, the one neutralizing the gyroscopic effect of the other. Instead of an undercarriage, the curved skids of the machine rested directly on a frontal bicycle-hub roller in a light truck with two flanged wheels, which fell away at the moment of launching. When the plane rose into the air under its own power, it vindicated to the full the Wrights' long-held belief that a motor-driven machine could be made to fly.

Although reporters were not present when the Flyer began to "soar like an eagle," one statement in the garbled *Washington Post* account was correct: a wind, averaging 21 m/hr, was blowing at the time. This was something of a paradox—for the chief enemy of early flyers was wind, of any velocity. A study of U.S. Weather Bureau reports had led the

Although the Washington Post *reported that their invention had "soared like an eagle" at Kitty Hawk, the Wright brothers' first plane never rose more than ten feet above the ground. The first flight, on December 17, 1903, covered a distance of 120 feet. (Air Force Museum)*

Wrights to believe that the desolate dunes of the North Carolina coast experienced constant winds of about 15 m/hr almost every day. This was the condition they had been seeking for their experiments with a glider. But the figure proved to be simply the daily average over a month; gales of 60 m/hr were sometimes encountered on a daily basis.

Circumstances, however, dictated a powered trial that December day despite the chill, blustery weather. It was very late in the season; the brothers had waited many weeks for the proper moment; and the night before, a biting north wind had sprung up, freezing the puddles from a recent rain and making work outdoors difficult. On the morning of December 17 the wind was clocked between 21 and 27 m/hr. When it failed to die down by ten thirty, the Wrights felt they should make an attempt regardless of conditions. Although Orville later looked back with amazement on their audacity in trying to fly for the first time under such circumstances, confidence in their calculations gave the bicycle makers from Ohio the courage they needed. Behind them were months of careful laboratory work, the painstaking compilation and checking of their own table of air pressures, and three years of experience in gliders, from which the design of their aeroplane had evolved. Much later their faith was to be commemorated by a national memorial on the top of Kill Devil Hill, the cornerstone of which was laid by the U.S. Government in 1928 on the twenty-fifth anniversary of the event. Nearby, the Spartan hangar and living quarters occupied by the brothers during their trials have been reconstructed; adjacent is a small museum containing a reproduction of their famous 1903 Flyer.

Sons of a clergyman—later a bishop—of the United Brethren Church, the Wright brothers had taken advantage of the newfound popularity of bicycles and enjoyed a moderate income from manufacturing, selling, and repairing them. Orville, the younger, born August 19, 1871, even engaged in bicycle racing. Both were on the quiet, scholarly side; both were intensely curious about all things mechanical. Wilbur, four years older than his brother, was born April 16, 1867. He was lean, blue eyed, clean shaven, prematurely bald, and taciturn, but with a wry sense of humor. When called on for a speech in France, he observed laconically: "The parrot is the only bird that talks, and he doesn't fly." The same qualities of mind and spirit were found in Orville. Slightly shorter and heavier than Wilbur, with the same eyes, and bearer of a short, thick moustache, he was the ideal teammate—working closely with his brother, sharing the thoughts and plans which eventually made them leaders of the world in aviation. By agreement neither ever married, and Orville broke with their sister, Katharine, who had shared their pact, when she wed Henry Haskell of the Kansas City *Star*. Wilbur is supposed to have said that neither of them had the means to support "a wife and a flying machine too."

As young men the brothers made up their minds to prove the possibility of flight when they read about the death of Otto Lilienthal, a German who led the world in gliding, in an accident in 1896. At that time Orville was recuperating from typhoid fever (an illness that later was to claim Wilbur's life). As soon as Orville was well, the "boys" embarked on what the neighbors liked to call their "crazy doings." In a letter to the Smithsonian dated May 30, 1899, asking for information and a list of books, Wilbur wrote:

I have been interested in the problem of mechanical and human flight ever since as a boy I constructed a number of bats of various sizes after the style of Cayley's and Penaud's machines [i.e., model helicopters]. My observations since have only convinced me more firmly that human flight is possible and practicable. It is only a question of knowledge and skill just as in all acrobatic feats. . . . I believe that simple flight at least is possible to man and that the experiments and investigations of a large number of independent workers will result in the accumulation of information and knowledge and skill which will finally lead to accomplished flight. . . . I am about to begin a systematic study of the subject in preparation for practical work to which I expect to devote what time I can spare from my regular business. . . . I am an enthusiast but not a crank in the sense that I have some pet theories as to the proper construction of a flying machine. I wish to avail myself of all that is already known and then if possible add my mite to help on the future worker who will attain final success.

The material supplied by the Smithsonian Institution showed that progress in the matter of mechanical flight since the futuristic dreams of Leonardo da Vinci had been anything but encouraging. Sir Hiram S. Maxim, a Maine-born American turned British subject, had dissipated part of the fortune he had earned from the invention of the Maxim gun on the construction of a large, multisurfaced, steam-powered air machine. In a hitherto unpublished letter headed Baldwyn's Park, Bexley, Kent, June 28, 1892, Maxim told a young graduate of Princeton who helped in the experiment—Edward Ringwood Hewitt, son of Abram S. Hewitt, the American industrialist and political leader:

The machine has already been running on the track without the sails [*sic*] and is all finished except putting on the sails. In a few days I shall be conducting experiments as regards the lifting power. I am not using the same steam generator but one in which there is a rapid forced circulation. I think you would like the boiler if you should see it.

But the apparatus was wrecked when it sprang the guide rail which restrained its lift.

In France, Clément Ader (1841–1925), on October 9, 1890, had become airborne for a distance of about fifty meters in a batlike affair with a 20-hp steam engine; but this was in no way a sustained or controlled flight. He had received a government subsidy in 1897 for a second machine, which had two 20-hp steam engines driving two feathery, four-bladed propellers of linen and rice paper that rotated in opposite directions. Most historians of aviation agree it never got off the ground—although it did give the word *avion* to the French language. Percy Sinclair Pilcher, Scottish engineer and prominent experimenter in the gliding field, had been killed shortly after Lilienthal. In Chicago, Octave Chanute, a renowned civil engineer of French extraction, had given up his glider exploits on the sandy dunes along the Indiana shore of Lake Michigan because of advancing age.

Nonetheless, in a letter to Chanute dated May 13, 1900, Wilbur observed that he had been "afflicted" for years "with the belief that flight is possible to man. My disease has increased in severity and I feel that it will soon cost me an increased amount of money if not my life. I have been trying to arrange my affairs in such a way that I can devote my entire time for a few months to experiment in this field."

This was the blueprint that the orderly and original-minded Wright brothers laid out for themselves. Their contribution to the science of aeronautics, however, was considerably more than a "mite"; for their 1903 machine incorporated basic features which—though the patents became subjected to litigation later on—were ultimately recognized as the rightful precursors of all modern methods of controlling an aeroplane in flight.

Business at the Wright Cycle Shop (now preserved in the Henry Ford Museum at Dearborn, Michigan) was soon mixed with the pleasure of exploring the problems of flight—how to maintain the equilibrium of a flying machine; where to place the center of gravity; how to calculate pressure of the air on a flat or curved surface; how to obtain directional control. Here, for example, grew the idea of twisting the tips of a wing ("warping") to preserve lateral balance—a necessity of successful flight that in one form or another would have to be incorporated in nearly every later aeroplane. While toying with a long cardboard box one day, Wilbur noticed that although the vertical end sides remained rigid, the top and bottom sides could be twisted to form a new set of angles at opposite ends. In much the same way, he had observed, gliding and soaring birds such as the buzzard or sea gull "regain their lateral balance . . . by a torsion of the tips of the wings." In Orville's words, "the basic idea was the adjustment of the wings to the right and left sides to different angles so as to secure different lifts on the opposite wings." Later the Wrights linked their wing-warping control with a rear rudder in order to counter-act the "drag" caused by warping—one of the greatest contributions they made to aeronautics.

The wing-warping principle, first applied in 1899 to a box kite with a fixed horizontal tail plane, was put to the test in a series of gliding experi-ments that began on the slopes of Kill Devil Hill, near Kitty Hawk, in 1900 and continued through the summers of 1901 and 1902. The earliest glider, constructed at a cost of some $15, was a biplane with a horizontal movable surface experimentally placed in the front for controlling climb and descent. Set at a positive angle so that the pressure of the air im-pinged on its underside, the horizontal elevator also provided a measure of inherent stability and relieved part of the load on the wings as well. This first glider had no rudder nor tail of any kind. It was followed by a second and larger glider, of the same general configuration. Having deter-mined that a wing camber (curve) of 1:12 was too pronounced, the Wrights reduced it to 1:19 in the second glider and greatly improved performance. The third and last glider, launched first with a fixed vertical surface and then with a rudder at the rear, was the result of an intensive program of research in Dayton, including the patient testing of innumer-able airfoil sections in a home-built wind tunnel and the thorough rework-ing of the brothers' aerodynamic theories. The final glider embodied the essentials of the Wrights' basic patent (No. 821, 393—applied for in 1903, though not granted until May 22, 1906): the increased angle of in-cidence at one wing tip and corresponding decrease at the other; the prac-tical technique of wing warping; and the simultaneous use of warping and rudder action to effect proper lateral control.

For the purposes of the gliding experiments, no more deserted spot could be found than the barren slopes of Kill Devil Hill. The area already had a reputation: in the eighteenth century the rolling dunes had been the haunt of rum smugglers, who traded the liquor for salt, fish, and tobacco. Some said that the name "Kill Devil" was derived from the strength of the rum. The origin of "Kitty Hawk" is more obscure. According to one picturesque legend, this resort of flocks of migrating geese attracted the Indians, who came to "kill a honk"—hence, originally, "Killy Honk."

But it was a steady breeze, not seclusion, that the Wrights were seek-ing. As Wilbur put it to Chanute, "I make no secret of my plans for the reason that I believe no financial profit will accrue to the inventor of the flying machine, and that only those who are willing to give as well as to receive suggestions can hope to link their names with the honor of its discovery. The problem is too great for one man alone and unaided to solve its secret." There was one further advantage in the location: the sands provided a relatively soft surface in case of a fall.

Power, stability, control; these were the principal problems confronting all pioneers in the effort to achieve successful flight. After executing more than a thousand glides in September and October of 1902—some in a wind of as high as 36 m/hr, some of more than 600 feet in length, and some lasting a minute or more—the brothers became convinced that they had learned enough about stability and control to go ahead. There remained the question of a motor sufficiently light in weight yet powerful enough to raise a machine from the ground and propel it through the air. The arrival of the internal combustion engine in the late 90's had made it possible to obtain the necessary power, but the bulk and weight of existing automotive engines rendered them wholly unsuitable for use in a flying machine. The motor the Wrights devised, a four-cylinder, water-cooled model without spark plugs that used a fuel injection system, was as much a contribution to the first free flight as the design of the aeroplane itself. The engine weighed 200 pounds, including accessories, and at 1200 rpm delivered all of 12 hp.

Following the historic flights at Kitty Hawk, the Wrights transferred their operations to a sixty-eight-acre cow pasture at Simms Station, eight miles out of Dayton, owned by a banker friend named Torrence Huffman. Now part of Wright-Patterson Air Force Base, the exact spot where their original hangar stood on the "flying field" is indicated by a marker. Here—on what was unquestionably the world's first aerodrome—the brothers continued experiments in 1904 and 1905 with improved versions of their famous Flyer. The airframe of the second of these models was disassembled and destroyed in 1905, while the third, in reconstructed form, is now preserved in a specially built hall in Dayton's Carillon Park. The Huffman tract was near a point in the landscape which became known as "the crossroads of twentieth-century transportation." Around the year 1910, six forms of travel could be seen from this point at once: the interurban trolley, the railroad, the Miami River, the Erie Canal, the automobile highway, and a Wright aeroplane.

Getting into the air at the Huffman field posed a difficulty not encountered at the Carolina proving grounds: here were no dependable winds to help lift the machine off its track. To assist the takeoff, an ingenious device was erected near the launching rail—a derrick with a falling weight which, through a system of ropes and pulleys, exerted a strong forward pull on the plane and sped it on its journey even in a dead calm. A total of 160 flights were made in this period, of up to 24⅕ miles in length and as long as 38 minutes 3 seconds in duration, including the first flight in a complete circle. This last achievement was a special triumph. The Wrights had suffered a series of tailspins and crackups before perfecting the necessary controls; in making a short turn, for instance, they found that a side-slip might be avoided by watching a short piece of string tied to the cross-bar beneath the front elevator. When the machine traveled straight ahead, the string trailed straight behind; when the machine slipped to one side, the string blew in the other direction. At this point, too, the brothers separated the warping and rudder controls, thus enabling their independent or combined operation in any desired degree—an arrangement that was to be modified only in later years. When the circle feat was finally accomplished, it was chronicled not in the newspapers but in the most improbable of scientific publications: *Gleanings in Bee Culture,* which still appears regularly in Medina, Ohio. Amos I. Root, then the editor of this folksy apiarists' journal, was properly amazed. In the issue of January 1905, he wrote:

Dear friends, I have a wonderful story to tell you—a story that, in some respects resembles the Arabian Nights. . . . God in his great mercy has permitted me to be, at least somewhat, instrumental in ushering in and introducing to the great wide world an invention that may outrank the electric cars, the automobiles and all the other methods of travel, and one which may fairly take a place beside the telephone and wireless telegraphy. Am I claiming a great deal? Well, I will tell my story and you will be the judge. . . .

To make the matter short, it was my privilege on September 20, 1904, to see the first successful trip of an airship without a balloon to sustain it, that the world has ever seen, that is, to turn the corners and come back to the starting point. . . .

Here, in this Ohio field, the Wrights taught not only themselves but others as well; here many of the pioneer aviators of this and other countries gained their wings. In recognition of the important advances made on this spot, several loads of its hallowed earth are today buried beneath a tall memorial shaft that overlooks the city of Dayton, commemorating in a bronze tablet "the courage, perseverance, and achievements" of the two men.

To an extent unbelievable today the Wrights continued to be prophets without honor in their own country. Idlers, neighbors, and passersby showed no comprehension of what they were watching; a group of newspapermen specially invited to a demonstration thought it a waste of time and dropped all interest when the motor of the Flyer failed to function and the wind was not favorable; scientific "proof" was offered to refute any claim to the possibility of flight. Abroad the Wrights were as controversial as are flying saucers today—they were "either flyers or liars," and in France a new word was coined: *bluffeurs.*

It is not surprising, therefore, that the Government of the United States failed to grasp the implications of an invention which, if its development had been properly encouraged from the start, would have put America far ahead in the air over the next decade. As early as January 18, 1905, the Wrights were sure they could offer something worthwhile to their country. Writing to their congressman, R. M. Nevin, the brothers pointed out what their flying machine could do:

It not only flies through the air at high speed, but it also lands without being wrecked. . . . The numerous flights . . . have made it quite certain that flying has been brought to a point where it can be made of great practical use in various ways, one of which is that of scouting and carrying messages in time of war.

Mr. Nevin forwarded the letter to the War Department, which turned it over to the Board of Ordnance and Fortification. The board, assuming that what the Wrights wanted was financial help, treated their offer as the production of a pair of cranks. Heavier-than-air flight was considered in the same category as perpetual motion; in answering Congressman Nevin the board's president, Major General G. L. Gillespie, wrote that first "the device must have been brought to the stage of practical operation without expense to the United States." Not once but twice in the year 1905 the military rejected the Wrights' proposal out of hand. In the face of such indifference, it was inevitable that the Wrights would respond to expressions of interest from abroad.

If the U.S. Army could see no use for a flying machine, France, which appeared to be heading for a clash with Germany over a "sphere of influence" in Morocco, might use one for scouting purposes in that uneasy territory. Through the efforts of Captain Ferdinand Ferber, himself a pioneer flyer, an option was obtained from the Wrights for the purchase of one aeroplane for the French army. As the brothers wrote prophetically to Captain Ferber on November 4, 1905:

When it becomes known that France is in possession of a practical flying machine other countries must at once avail themselves of our scientific discoveries and practical experience. With Russia and Austria-Hungary in their present troubled condition and the German Emperor in a truculent mood, a spark may produce an explosion at any minute. No government dare take the risk of waiting to develop practical flying machines independently. To be even one year behind other governments might result in losses compared with which the modest amount we shall ask for our invention would be insignificant. [France] . . . may wish to avail itself of our discoveries, partly to supplement its own work, or, perhaps, partly to accurately inform itself of the state of the art as it will exist in those countries which buy the secrets of our motor machine.

Under the circumstances, the Wrights continued, they would consent to "reduce" the price of their aeroplane to the French government to 1,000,000 francs, or $200,000. Ferber had not been able to obtain a commitment from his government as to the price it would be willing to pay —and the Wrights did not mention their original offer from which the price of 1,000,000 francs had been reduced. In any case, since no one in France had yet seen the Wrights fly, War Minister Eugène Etienne thought it advisable to send an investigating commission to Dayton before signing a contract. While the advantages of an airworthy scouting craft were obvious, the French were not sure that a practicable machine existed except in the inventive minds of the Americans. However, while the commission was still in Dayton, the war threat was terminated (April 7, 1906) by the Conference of Algeciras. Settlement of the dispute made the need for a scouting plane less urgent, and stiffer terms were imposed by the French in amendments to the contract. Skepticism continued to prevail; the commission was recalled, and the negotiations came to nothing. At about the same time British feelers were put out, but these seemed to be based more on a desire to find out what other countries were doing than on any serious intention to acquire a flying machine for military use.

When the Wrights were turned down by the U.S. War Department for the second time in October 1905, they withdrew from active flying, grounding themselves for a period that was to last two and a half years. Although this no doubt was a setback to the development of aviation, the time was not technically wasted. The brothers built new engines of improved capability and continued to develop their Flyer. To obviate the need for lying flat on the lower wing surface, they evolved a system of control levers, to be operated by hand from a sitting position, and added a passenger seat alongside; but they retained the skid undercarriage and the derrick-and-rail, or "guillotine," type of launching arrangement. The machines produced on these lines between 1907 and 1909 later became familiar as the standard Wright Model A biplane. These planes represented not only the culmination of the Wrights' pioneering experiments but also the type of aeroplane first seen in public demonstrations.

Such a machine was taken to France by the brothers in the summer of 1907 (though, for the time being, left unpacked) in anticipation of possible trials either for the French government or for a private syndicate that might acquire the manufacturing and licensing rights for foreign countries. Before taking this step, however, the Wrights renewed correspondence with the Board of Ordnance and Fortification because of interest displayed by President Theodore Roosevelt. On this occasion they made a formal proposal to sell one or more of their machines to the U.S. Government—the price to be $100,000 for the first machine, with others to be furnished at a reasonable margin over the cost of manufacture. No payment was to be made until a trial flight had proved compliance with all contract specifications. When the board again demurred, citing a requirement that such a sum would first have to be appropriated by Congress, the brothers made clear that if price was the only obstacle the matter could doubtless be adjusted. Privately the brothers agreed between themselves that their final asking price for the sale of an aeroplane to the U.S. Government would be $25,000.

The visit to Europe brought upon the stage an American businessman, Hart O. Berg, who was to play a principal role in the Wrights' affairs as their European representative. Berg was an associate of the New York firm of Charles R. Flint & Company, bankers and promoters, who saw financial possibilities in the sale of Wright biplanes—if not in the United

States, then on the European Continent. Berg had been active in sales abroad of American electric automobiles; he had also dealt with the Russian and other foreign governments in behalf of Simon Lake, inventor of the submarine. In addition to helping the Wrights in France, he steered them into discussions with German government officials in Berlin. Berg soon became one of the earliest and most enthusiastic go-betweens in the field of flying. With his help, licenses to make Wright-type machines were granted to a syndicate in France, as well as to the German and Italian companies in 1909, and finally to the British in 1913. It was arranged that the Bariquand et Marre motor works—building, under license, faithful copies of the Wright engine—would supply the standard motive equipment for French-made Wrights; these engines were even imported for use in Wright aeroplanes in the United States. Strangely enough, despite the success of their manufacture, they were never put to use in other aircraft in Europe.

During the Wright brothers' pause in aerial activity, another inventive American became involved in flying. Glenn Hammond Curtiss, born May 21, 1878, in Hammondsport, New York, was also a bicycle maker and idea man. His taste for speed led him into the business of building and racing motorcycles, on one of which on January 29, 1904, at Ormond Beach, Florida, he streaked 10 miles in 8 minutes 54⅖ seconds—a world's record that stood for the next seven years.

Captain Thomas S. Baldwin (1864–1923), California's pioneer dirigible balloonist, brought Curtiss into the flying fold by ordering one of the lightweight Curtiss motorcycle motors for his nonrigid, gas-inflated, cigar-shaped airship. Baldwin built more than a dozen dirigibles, all equipped with Curtiss motors, and in the summer of 1908 delivered a specially constructed military model to the Aeronautical Division of the U.S. Army Signal Corps in Washington. With Baldwin operating the controls from the rear of the skeleton framework and Curtiss, balanced up forward, looking after the motor, the craft easily met government specifications in a two-hour flight over the Virginia countryside. Apparently there was little difficulty in convincing the War Department that lighter-than-air vehicles had a military future. Balloons had been used for observation in the Civil War and as an escape device during the Franco-Prussian War, while dirigibles, in these early years of the twentieth century, had already caught the eyes of military officials in France, Germany, and England.

Curtiss's work on motors attracted the attention of the many-sided scientist Dr. Alexander Graham Bell, educator of the deaf and inventor of the telephone. Bell, born in Edinburgh, Scotland, in 1847, was nationalized as an American citizen in 1882. He was an imposing figure, with dark, penetrating eyes and bushy white hair and beard. His wife Mabel (daughter of the railroad magnate and banker Gardiner Greene Hubbard), although deaf from childhood, was an exceptionally interested helper in his manifold undertakings. Curtiss and Bell first met in New

York in 1905, at which time Bell invited Curtiss to visit him at his summer home, Beinn Bhreagh (Gaelic for "Lovely Mountain"), near Baddeck on Cape Breton Island, Nova Scotia. In that cool, remote retreat
among the rocks and pines, Bell had been conducting a series of experiments with tetrahedral kites—four-sided, lightweight aluminum frames
covered with silk—one of which, a large and relatively strong model, possessed great inherent stability. Bell was anxious to attach one of the Curtiss motors to it as part of his studies in aerodynamic lift, propulsion, and
control, for he had set his sights on the construction of a machine that
would fly even before the Wrights had taken off at Kitty Hawk. In January 1903, for example, Bell was quoted by the Boston *Transcript* as hy-
pothesizing that "an aeroplane kite could carry the weight of a motor and
a man." Realization of this exploit would be only one step short of the
goal of free flight.

Bell had gathered around him at Baddeck a group of bright young men,
including two recent graduates in mechanical engineering of the University of Toronto: Frederick Walker ("Casey") Baldwin (no relation to
the balloonist) and John A. D. McCurdy, son of an inventor, who was to
mature into one of America's foremost aviators. Bell tried to persuade
the Canadian government to send an observer to join the group, but official circles were not interested. Lieutenant Thomas E. Selfridge of the
U.S. Army was, however, an active participant. In the pleasant, informal
surroundings of Bell's spacious estate and under his jovial and sympathetic direction, a wide-ranging discussion of the theories and principles of
aerial locomotion took place. As a result and at Mrs. Bell's suggestion, an

The first flying machine developed by the Bell group was the Red Wing, a biplane with the propeller located behind the wing. It is shown here with pilot "Casey" Baldwin before its flight at Lake Keuka, New York, on March 12, 1908. (Air Force Museum)

The Red Wing flew for a total of 318 feet 11 inches (at right). But its first flight was also its last. Having no lateral stability, it rolled over on its side, crashed, and was completely demolished (bottom). (Air Force Museum)

Aerial Experiment Association was formed. Bell was chairman, without salary; Baldwin was chief engineer, and McCurdy assistant engineer and treasurer, at $1000 a year each; Selfridge was secretary, at $5000 a year. When Curtiss joined the group in the summer of 1907, he became director of experiments and chief executive officer; he received $5000 a year while at the scene of operations and thus absent from his regular business, and half pay at other times. Mrs. Bell (who was independently wealthy), having provided a name for the organization, supplied funds as well—to the extent of $35,000. The official purpose of the association was "to build a practical aeroplane which will carry a man and be driven through the air on its own power."

Bell's investigations were influenced to a substantial degree by the cellular form of kite construction developed in 1892 by Lawrence Hargrave, an Australian, who firmly believed in the applicability of kite experimentation to powered flight. After the Bell group had conducted trials at Baddeck with motor-driven tetrahedral kites and with aerial propellers mounted on boats, the experimenters moved to the Curtiss workshop at Hammondsport, on the shores of Lake Keuka in the Finger Lakes district of New York. Here a biplane glider was built and tested on the snow-covered slopes around the lake until the members of the association felt ready for tryouts with a motor-driven machine.

The vehicle was the Red Wing—a bird of bright plumage. While incorporating the suggestions of all the experimenters, it was mostly the creation of Lieutenant Selfridge. Selfridge had approached the subject of powered flight after a thorough study of gliders. To him was entrusted the responsibility of building this frail prototype of a pusher biplane in accordance with data assembled by the association. (A "pusher" has its propeller behind the wing, in contrast to a "tractor," which has its propeller at the front.) Selfridge, the first military person in the world to fly an aeroplane, was not only thorough but courageous in his attack on the problems of flight; he was the tireless leader of the group in its penetration of the aeronautical mysteries. He was also a careful and methodical man, confident that he was taking no chances. When, as Curtiss later related, a special agent for a life insurance company visited the field where they were practicing and laughingly warned Selfridge to take care, "or we will need a bed for you in the hospital of which I am a trustee," the answer was perfectly sincere: "Oh, I am careful, all right." It was through no fault of his own that a career of such promise was soon to be cut short. In 1908 Selfridge was killed in the crash of a plane piloted by Orville Wright.

On a bitterly cold March 12, 1908 (temperatures did not seem to matter much to the hardy pioneers), the Red Wing, piloted by Casey Baldwin, sped over the icy surface of the lake on runners, bounded into the air, and actually flew for a distance of 318 feet 11 inches. Being virtually uncontrollable since it lacked any stabilizing device, it flipped over on one side and crashed. However, disregarding the practically unpublicized flights of the Wright brothers, this was the first time that an aeroplane was flown publicly in America.

The Red Wing was followed in a few weeks by a resplendent White Wing, designed by Baldwin. This model, because the ice had melted, was put on a tricycle undercarriage and taken for trials to an abandoned race-track known as Stony Brook Farm. It was soon apparent that to get the White Wing into the air was one thing, but to get back down without wrecking the machine was quite another. Smash followed smash in discouraging succession—fortunately with no injuries save to the feelings of the operator. "It seemed one day that the limit of hard luck had been reached," wrote Curtiss of these first ventures, "when, after a brief flight and a somewhat rough landing, the machine folded up and sank down on its side, like a wounded bird, just as we were feeling pretty good over a successful landing without breakage." The only way to learn was the hard way: by trial and repair, by study of stresses and strains, by provisional changes in details of construction. But on May 22 the White Wing, with Curtiss at the controls, flew a distance of 1017 feet in 19 seconds and actually landed intact in a ploughed field outside the old racetrack. It was cause for elation—and for the prompt construction, under Curtiss's direction, of a bigger, better, prize-winning plane: the June Bug.

Clearly enough a flying machine could not fly properly without balancing itself like a bird. On Alexander Graham Bell's recommendation hinged surfaces were built into the wing tips to serve as lateral rudders, or "ailerons"—operated by the instinctive movement of the aviator through a shoulder harness—which performed the same function as the Wrights' technique of wing warping. By leaning to the side of the high wing (a natural motion), the pilot could raise the aileron—which had the effect of lowering the wing. Improvement was immediately noticeable. Although the incorporation of ailerons in the Curtiss machine was to lead to costly and acrimonious lawsuits, the idea evolved entirely independently of the Wright brothers' theories. When Orville Wright, in a letter to Curtiss dated July 20, 1908, fired the opening gun with a charge of patent infringement, Bell, greatly disturbed, was at pains to make clear that his suggestion regarding movable wing tips had been submitted without prior knowledge of the Wright invention. Technically almost anyone who built a workable aeroplane was guilty of infringement; for the Wright patent to a considerable degree covered the whole range of flyable machines. But the Wright-Curtiss battle was particularly venomous and protracted. The case was not settled until January 1914, when the court ruled that wing warping and aileron action were essentially similar. Ironically the Wright system of control was to be abandoned within a decade; the Aerial Experiment Association's principle of the aileron, patented on December 5, 1911, came to be universally adopted.

As an award for the first public flight of 1 kilometer (or ⅝ mile) in a straight line, the magazine *Scientific American* had offered an impressive silver trophy—a globe supported by a pedestal, surmounted by an eagle,

Equipped with hinged ailerons on its wing tips, the June Bug, flown by Glenn H. Curtiss, won the Scientific American *trophy on July 4, 1908, for successfully covering 1 kilometer in the air.* (Air Force Museum)

and surrounded at the base by winged horses. The rules specified that the first winner for three years in succession could keep the trophy permanently. The conditions of the contest were to vary with the progress of aviation. On the Fourth of July, 1908, the June Bug made its first bid to win the trophy. With flags, fireworks, and all-day picnics to celebrate the occasion, the citizens of Hammondsport were joined by a small band of enthusiasts from the Aero Club of America—an organization formed in New York in 1905 by Augustus Post, balloon enthusiast and motorist, and a number of kindred spirits. They were accompanied by a similar group from Washington. Such clubs were few in number and vague in their programs, but the handful of individuals who composed them were earnestly air minded. Thus there were witnesses aplenty to make the flight official; even the trees held spectators. None, however, were more excited than the founders of the Aerial Experiment Association. The June Bug's engine had been tuned up with meticulous care, and an attempt had been made to render the wing surfaces airtight by coating the silken fabric with varnish—an early example of the use of aircraft "dope." If anything was likely to defeat the effort, it would be the weather.

The scientific study of air currents was then in its infancy. A year or two earlier A. Lawrence Rotch, director of the Blue Hills Observatory, had noted in his book *The Conquest of the Air:* "At night . . . because there are no ascending currents, the wind is much steadier than in the daytime, making night the most favorable time for aerial navigation of all kinds. . . ." Obviously the night hours were hardly a practical time for the early birdmen to take wing. The next most favorable time for successful flight seemed to be in the calm that generally accompanies the sunrise or sunset, when the turbulence caused by thermal currents during the heat of the day could be avoided. This rule of thumb had governed the plans for testing each of the three previous association machines and would con-

*Built at Hammonds-
port, New York, the
Silver Dart was
transported to Canada
for the first heavier-
than-air flight in that
country, on Febru-
ary 23, 1909. John
A. D. McCurdy
was the pilot.
(© H. M. Benner
and National
Geographic Society)*

tinue for years to be a cardinal condition for aviators who took no chances.

At five o'clock on the morning of that Glorious Fourth there were clouds and wind; at noon it was raining; but late in the afternoon conditions began to improve. About seven o'clock, in gathering dusk and against a dramatic background of green-clad hills and inky black clouds, Glenn Curtiss handily captured the trophy. With a perfectly functioning motor supporting him, he flew past the flag marking the end of the course and came down in a meadow more than a mile from the starting point. Only a forbidding row of trees and the frightening specter of making a turn in the air prevented him from going farther. The speed of this flight—the first to be officially recorded in America—was 39 m/hr.

Next and last of the Aerial Experiment Association's pioneer aeroplanes was the shimmering Silver Dart, designed by McCurdy, which made its debut on December 12, 1908. The Silver Dart did not spend the winter at Hammondsport but was transported to Baddeck, where on February 23, 1909, it was flown for a distance of ¾ mile by McCurdy—the first flight of a heavier-than-air machine in Canada. In some 200 subsequent flights this craft was estimated to have covered a total of 1000 miles.

With the achievements of the Silver Dart the activities of the Aerial Experiment Association came to an end. Its point proven, the society was dissolved. Bell went back to his kites and died in 1922; his simple grave, marked by a boulder with a plaque atop the hill overlooking his summer home and the lakes he loved so well, fittingly commemorates the twenty-five years of his life that he consecrated to the development of flying.

McCurdy and Baldwin became the fathers of Canadian aviation, helping to found the first Canadian aircraft manufacturer—the short-lived Canadian Aerodrome Company. One of the five machines it produced was tested for the Canadian army at Petawawa, Ontario, in the summer of 1909. With McCurdy at the controls it landed in a rough field—quite in contrast to the smooth ice at Baddeck—and disintegrated. McCurdy escaped with a broken nose (the only injury he ever received in his flying career), while the army took the accident as proof that the aeroplane had no practical use in warfare.

Curtiss went on to develop the compact pusher biplane that was to set many records in the years to come. And when Canadians failed to summon enthusiasm over the prospects of powered flight, Baldwin returned to the United States for new experiments with a Curtiss-designed hydroaeroplane, while McCurdy became the chief pilot for Curtiss—the first to take a "flying boat" into the air and the first (in August 1910 at Sheepshead Bay, N.Y.) to transmit a wireless message from a plane.

CHAPTER THREE

Awakening in France

Nothing comparable with the advances made by the Wrights had been similarly achieved in Europe, and for the first six years of the twentieth century theirs was the only name readily associated with the art of flight. "One of the strangest mysteries in aeronautical history is the confusion and procrastination which enveloped the Europeans and seriously delayed the arrival of the aeroplane in the world at large," wrote Charles Harvard Gibbs-Smith in a contribution by Britain's Science Museum to the centennial observance of the Royal Aeronautical Society in 1966. ". . . despite the fact that all the necessary clues to success were in the hands of the Europeans by the end of 1903, it was not until the end of 1906 that the first tentative powered flights were at last made in Europe."

But the first reports of glider and flying-machine trials in America had kindled a latent spark in France. Octave Chanute came to Paris in 1903, lectured before the Aéro Club de France on the lessons learned at Kitty Hawk, and published a series of articles on gliding in French magazines. Had it not been for his persuasive arguments that man was on the threshold of mechanical flight, it is probable that the Paris club would have continued to concentrate on balloons and dirigibles to the virtual neglect of other means of aerial travel. A small group of members, however, led by Ernest Archdeacon, an affluent Paris lawyer and auto enthusiast, were determined to advance the cause of powered planes. The best way to start seemed to be through the promotion of glider prizes and contests. Archdeacon, an ardent proponent of all kinds of rapid transportation, whose staccato, high-pitched voice often dominated the debates in the club, led off with a subscription of 3000 francs. He was supported by other wealthy members, including the noted aeronaut Count Henri de la Vaulx (one of the founders of the Aéro Club), whose interest was turning from lighter-than-air to mechanical flight, and Henri Deutsch de la Meurthe—a powerful, bearded industrialist and oil magnate, whose name soon became a synonym for substantial monetary awards. It was in fact through the generosity of such farsighted men, patriotically determined to send a

French-made machine into the air and thereby capture laurels for France, that technology received its greatest impetus. To this factor must be added the stimulus of competitions and aerial meets staged under the auspices of an enterprising press—often resulting from rivalry among newspaper tycoons. The lure of valuable trophies or prizes in cash spurred the progress of powered flight over the next decade.

Archdeacon substantiated his interest by ordering the construction of a glider along the lines (as then known) of the 1902 Wright model. To test his glider he engaged a young graduate of an architectural school at Lyons, Gabriel Voisin, who served as his "engineer" at the munificent salary of 190 francs a month. Archdeacon soon commissioned Voisin to build a second glider, thus laying the foundation for what was to become the famous firm of Voisin Frères.

Gabriel Voisin was reared in a family long connected with industry. At an early age he had shown a mechanical bent. After his introduction to Archdeacon in 1903, Gabriel decided to give up everything else for aviation. Pulling on an old jersey sweater, he experimented first with gliders on the dunes of Berck-sur-mer, in 1905. Then he progressed to pontoon-fitted gliders towed by motorboat, at Billancourt on the Seine and later on the Lake of Geneva. Soon he began constructing and piloting powered planes. Gabriel was a handsome man, with the piercing eyes of a sparrow hawk and a black moustache. His brother Charles, no less good looking, with whom he had built and flown kites as a boy, presently joined him to form a second notable team of brothers in the advance of aviation. Charles served as the businessman of the firm of Voisin Frères until he was killed in an automobile collision near Belleville in 1912.

Even more active than the Voisins during the gestative period of French flying was Louis Ferdinand Ferber (born in 1862), a heavyset artillery corps captain with handlebar moustache who—like the Wrights and at about the same time—derived inspiration from the experiments of Lilienthal. The German glider expert was considered an acrobat by his countrymen, a parachutist by the French—so little were his researches understood. Only by serious students of flight were his efforts recognized as representing the beginning of knowledge in the field of aviation. Following his lead, Ferber, as early as December 7, 1901, was conducting glider experiments in the south of France. Ferber's philosophy was summed up in the motto: "To construct a flying machine is nothing; to build one is not much; to test it is everything."

Conveniently granted a three-year leave from the army, Ferber arranged to examine the correspondence on the Wright experiments between the American brothers and Chanute; then, entirely at his own expense, he tried out a biplane glider in the mountainous region of Breuil, above Nice, in 1902, and on the beach at Conquet in Finistère in 1903. Returning to Nice, he proceeded to install a motor in his machine and then suspend it from the arm of a lofty tower, to cut circles in merry-go-round fashion. Ferber followed avidly all reports about the Wrights, absorbing every

scrap of data available and steadfastly proclaiming his belief in their accomplishments at a time when skepticism reigned supreme. It was, of
course, Ferber's insistent prodding of the French Ministry of War that
led to its dickering for a Wright flying machine in 1906 and 1907. And it
was Ferber whom the Wrights suspected (perhaps unjustly) of permitting
those negotiations to lapse as soon as the Moroccan crisis had simmered
down—and then using the knowledge he had acquired to build a biplane
of his own, a tractor, with which he finally rose in free flight at Issy-les-
Moulineaux on the outskirts of Paris, on July 25, 1908. Actually the machine was put together by the Société Antoinette and powered by a 50-hp
Antoinette motor; however, it made only brief, straight-line hops.

Between 1904 and 1905 Ferber wrote a number of basic scientific bro-
chures, the titles of which aptly indicate the methodical manner in which
the early pioneers added to the accumulation of knowledge: *Progress in
Aviation by Gliding; Calculations; Step by Step, Jump by Jump, Flight by
Flight*. In his last treatise, *Aviation* (published in 1908), Ferber virtually
admitted that by successive stages he had re-created the Wright invention,
in somewhat the same fashion that paleontologists reconstruct an entire
prehistoric animal on the availability of a few fossil bones. Speaking of
Wilbur, he wrote: "Without this man I would be nothing . . . without
him my experiments would not have taken place." At the same time
Ferber recorded his disappointment with his military superiors—who
treated him as a harmless fool (*"un doux illuminé"*)—for not purchasing
a flying machine for the army. Ferber claimed to be the first in the world
to appreciate the achievement of powered flight as well as what it could
mean for France, particularly in a military sense. Whatever the facts may
have been, there is no doubt that in his perception he was well in advance
of his countrymen—a tireless gadfly of French aviation and a pioneer in
Europe like the Wrights in America.

Ferber had few rivals—at any rate none with his sense of discipline and
step-by-step thoroughness. (Looking backward, the French later were to
reproach themselves for "inexcusable torpor.") Despite all the published
data, the discussion, the intriguing reports about the Wrights, hardly any
other Europeans managed to get themselves into the air. True, there were
tentative stirrings, as from birds in a nest, to show that the moment to
take flight had almost arrived. In Denmark Jacob Christian H. Ellehammer (1871–1946), an engineer and builder of several successful
gliders, started construction in 1905 of a monoplane with a semicylindri-
cal wing, equipped with an engine rated at 18 hp. In January 1906 he
tested the machine on a circular track, tethered to a tall pole—but it did
not fly. In September, on the little island of Lindholm, Ellehammer converted his monoplane to a semi-biplane with tractor propeller but no
rudder; its pilot's seat, as in the previous model, swung like a pendulum to
maintain longitudinal stability. In the presence of a naval lieutenant
named Uelditz, acting informally as observer, the Danish birdman was
reported to have achieved a tethered flight of 42 meters at an altitude of

½ meter on September 12, 1906. Within the next two years he was to do somewhat better with a triplane. A 30-inch scale model of the successful 1906 machine was presented to the Smithsonian Institution in 1962 by the late Danish ambassador to Washington, Henrik Kauffman.

Also in 1906, an expatriate Rumanian lawyer and engineer living in France, Trajan Vuia, built a single-winged apparatus driven by a 25-hp Serpollet motor that ran on carbonic acid gas. It had a comfortable basket seat mounted on a short platform with four rubber-tired wheels, and a propeller (designed by Victor Tatin, an early aeronautical engineer) that looked like the arms of a windmill. On March 18 at the French town of Montesson, this primitive monoplane—the first in the world of its type—was credited with several short hops. Vuia's machine qualified sufficiently as a famous first to find a resting place, as the gift of the inventor, in France's Musée de l'Air at Chalais-Meudon—a vast Valhalla of antique aircraft and motors that seems to hold the essence of the heroic pioneer spirit.

Not until October of 1906, however—nearly three years after Kitty Hawk—was the record of the first flight in Europe of a heavier-than-air machine formally placed on the books, authenticated by the presence of official observers. Alberto Santos-Dumont, scion of a rich Brazilian coffee planter, was born July 20, 1873, in the district of João Aires, state of Minas Gerais. He came to Paris at the age of eighteen, filled with youthful enthusiasm for the twin novelties of balloons and automobiles. Santos-Dumont had dreamed of flying ever since he saw a balloon ascension at São Paulo; but, as he later admitted in his book *My Airships,* "these imaginings" were kept to himself. "In those days, in Brazil, to talk of inventing a flying-machine or a dirigible balloon, would have been to stamp one's self as unbalanced and visionary. Spherical balloonists were looked on as daring professionals not differing greatly from acrobats; and for the son of a planter to dream of emulating them would have been almost a social sin."

With leisure and money—two powerful assets in probing the secrets of flight—Santos-Dumont could afford to indulge his fancies. His "steerable balloons," constructed early in the new century, became the toast of sophisticated Paris. Fame was further assured when, on October 19, 1901, the adventurous Alberto won a prize of 100,000 francs—offered by the "angel" of aeronautics Deutsch de la Meurthe for the first airship to fly from Saint-Cloud around the Eiffel Tower and back. Half of these proceeds Santos-Dumont magnanimously gave to the Paris poor; the rest he divided among his mechanics.

Santos-Dumont was of the ideal type for an aeronaut. Slight of stature, nimble as a cat, his restless scientific mind and cool daring led to triumphant exploits as well as near disasters. He was impeccable in dress and affected a trim moustache. Wearing a broad-brimmed, floppy panama hat and usually a high, stiff collar, often sporting a carnation in his buttonhole, he charmed all who encountered him with a ready smile and serious

Nearly three years after Kitty Hawk, Alberto Santos-Dumont, the son of a Brazilian coffee planter, made the first recorded flight of a heavier-than-air machine in Europe. His biplane flew 50 meters at the Bagatelle cavalry grounds in Paris.

dark brown eyes. There was no doubt in anyone's mind that he deserved the tribute paid to him by an authority on aerial navigation after his trip around the Eiffel Tower: "In one short morning the question of aerial flight has been solved and solved for good."

Then in 1904 and 1905, curious about other ways of flying, Santos-Dumont carried out some novel experiments with a glider. In a machine equipped with pontoons, built for him by Voisin Frères, he was towed up the river Seine by a fast motorboat and then lifted through the air like a kite. By 1906 Santos-Dumont had diverted his talents to the construction of a powered plane. His Model XIV *bis* (so designated because he had tested it suspended beneath his airship No. 14) was colloquially called the Bird of Prey. But it looked more like a duck with outstretched neck: twin box kites set together at an angle, with a smaller one serving as elevator and rudder located at the end of a long framework in front. This unconventional affair, which appeared to fly backward, gave the nickname *canard* ("duck") to machines of similar type. The Santos-Dumont *canard* was the first (and perhaps the only) aircraft in history in which the pilot manipulated the controls while standing up—a position scarcely conducive to poise or comfort—rather than prone or seated.

As was so often the case in early experimentation, a suitably light motor—and one, above all, that would function uninterruptedly for a reasonable length of time—was difficult to find. At length an eight-cylinder Antoinette engine developing a fairly satisfactory 50 hp was installed at the biplane's tailless rear and connected with a four-bladed metal propeller. Supported on two bicycle wheels, the machine staggered into the air at the Bagatelle cavalry grounds in Paris on October 23, 1906, and flew approximately 50 meters before an excited group of formal observers from the Aéro Club de France—thereby winning the Archdeacon prize of 3000 francs for the first flyer to cover 25 meters. Santos-Dumont bettered this performance twice on November 12, first with 82.6 meters in 7⅕ seconds and then with a wavering flight of 220 meters in 21⅖ seconds, gaining a further reward from the Aéro Club of 1500 francs. The instability of the plane's design, however, led to its abandonment the following year, and Santos-Dumont turned his attention to a totally different model.

These were puny efforts compared with what the Wrights had already accomplished. But the Brazilian was to contribute more substantially to progress, beginning late in 1907, with the world's first light aeroplane—a bamboo monoplane, likened to a butterfly, which was fitted with a 20-hp Dutheil-Chalmers motor and a huge wooden propeller. The aviator sat beneath the wing, only a foot or so off the ground. Later known as the Demoiselle, this model seemed fearsomely frail; but its very lightness may have been the reason none of its pilots ever met with a fatal accident.

Here, with the Bird of Prey and the Demoiselle, was the true genesis of mechanical flight in France—and indeed on all the Continent. As its exponent, Santos-Dumont received an acclaim out of all proportion to his achievement, as set against the records already established at Kitty Hawk and Dayton but ignored or discounted by most of the world. To certify such records an official body, the Fédération Aéronautique Internationale, had been set up in Paris in October 1905 by representatives from Belgium, France, Germany, Great Britain, Italy, Spain, Switzerland, and the United States. But since no official observer was present during the American experiments, the success of Santos-Dumont was long afterward hailed by the uninformed as the first authenticated instance of power-driven flight. The United States had given birth to the aeroplane and technologically was far in the lead—but few people were aware of that fact. Even fewer were stirred by the thought that the era of flying was at hand.

Up to this point, flights were not only of short duration but were executed in a single direction. Always excepting the experiments of the Wrights, the early birds were glad to get off the ground, proceed in a straight line, and descend (if they could) without crashing. Gradually the straight line lengthened and rudimentary attempts to change course were made, none of which convinced the skeptics that a stable flying vehicle, obedient to the will of the pilot, would ever be practicable. For steering a heavier-than-air craft around a closed course of 1 kilometer, the Archdeacon–de la Meurthe prize of 50,000 francs was offered. Representing a considerable sum in those golden days of hard currency, the prize stood as an inducement to achieve what no one believed the Wrights had accomplished three years earlier.

Henry Farman, a small, energetic man who had abandoned an ambition to be an artist in favor of motorcars and aviation, set about winning that prize. Born in Paris on May 26, 1874, Henry (christened Harry Edgar Mudford Farman) was the second of the three sons of Thomas Frederick Farman, a wealthy Englishman who had come to Paris as correspondent for the London *Standard* and stayed on the job for thirty years. Thomas was an ardent convert to the pastime of bicycling and could often be seen riding around town in a top hat. Although both his parents were English and his birth was registered at the British Consulate in Paris, Henry Farman spoke the English language haltingly. He often spelled his name in the French fashion—Henri—and eventually received French citizenship by decree. The three Farman brothers, Henry, Dick, and Maurice, acquired their father's passion for cycling and became adepts at the sport; Henry and the younger Maurice were an invincible team in tandem races.

Unlike the Wrights, the Farman brothers did not build a business out of the bicycle boom. The budding motorcar industry attracted them much more; and Henry, in partnership with Dick, formed an agency for Panhard-Levassor, Renault, and Delaunay-Belleville automobiles. Motorcar racing had a special appeal for Henry, and he captured many prizes. After a near-fatal mishap in the Gordon Bennett road race elimination

Automobile racer Henry Farman turned to flying planes after a narrow escape in a road race. He is shown here beside the tail of his first machine, built for him by the Voisin brothers.

trials in 1905, however, he turned his attention to the "safer" sport of flying—and found it even more fascinating than motorcar racing. After first experimenting with model aeroplanes of various sizes, he began practice trials in 1907 with a homemade biplane glider on the sandhills of the English Channel town of Le Touquet. The next step was to order a powered machine.

At this date in Europe only the brothers Gabriel and Charles Voisin were in a position to construct and deliver a motor-driven aeroplane. The house of Voisin Frères believed firmly in the concept of inherent stability—which in this case meant no control at all in a bank or a roll, no real ability to maneuver except up or down. The Voisin pusher biplane, with forward elevator, side panels (or "curtains") to aid lateral balance, and a cellular tail enclosing the rudder, was in effect a steerable box kite. It was easy to fly in a flat calm but lacked any device to keep the wings horizontal, as in the Wright aircraft. The only way a turn could be made was to yaw the plane around and then, when one wing rose and the other fell in consequence, to yaw it the other way—a dubious procedure from the viewpoint of efficiency, but one which was influential in the basic design of other inherently stable "box kite" biplanes. The Voisin model, equipped with that mainstay of early aviation the 50-hp Antoinette motor, was Europe's first standardized practical aeroplane, and with some modifications was soon to become a favorite of many pilots on the Continent.

By October 1907, Voisin Frères had built two of their heavy planes on a commercial basis: one for Henry Farman and one for Farman's friend and rival, the sculptor Léon Delagrange. Delagrange, born at Orléans in 1873, was the son of a prominent industrialist engaged in the manufacture of cloth and thread. From an early age Léon showed a talent in sculpture and an interest in sports, preferring these endeavors to business. He exhibited regularly at the Salon sponsored by the Société des Artistes Français, where he was the recipient of several medals. In 1905 he became interested in the motorboat-glider experiments of Gabriel Voisin and commissioned Voisin Frères to build him a powered aeroplane. Delivery of this machine was made on March 15, 1907, at Bagatelle by Charles Voisin, who tested it with a hop jump of 60 meters.

Delagrange in his early experiments would cart his machine by night to the Bois de Boulogne or the park at Vincennes; then, after the inevitable mishap during the inconclusive trials performed the next day, he would take the pieces back to the shop for repairs. But Henry Farman started a new custom: he obtained permission to practice on the military parade grounds at Issy-les-Moulineaux. Situated on the left bank of the Seine just outside one of the old gates of Paris, with the Eiffel Tower seen in the distance, Issy was a natural flying facility. The flat, if dusty and sometimes muddy, field was bordered on the north by the city's walled fortifications, the grassy tops of which provided unrestricted space for spectators; to the west, casting shadows in the setting sun, stood the huge hangars of the Société Astra and Clément-Bayard dirigibles—whose yellow bulks occasionally floated over the scene. Two or three tall factory chimneys silhouetted against the sky provided a built-in wind sock: if the smoke rose straight upward, conditions were perfect; if it blew slantwise, it showed the direction in which to land; if it streamed straight outward, flying was next to impossible. Gradually Issy—the first actual aerodrome in Europe—became ringed with wooden sheds as additional flyers followed Farman there. With its easy accessibility by the subway system, the field afforded Parisians a convenient gallery from which to witness the start of many famous flights and races, enabling them to observe the unfolding of the drama of aviation at first hand over the years. The facility is still used—as a heliport.

Here Farman, as well as Delagrange, practiced regularly; and here Farman made history by taking his Voisin biplane around a closed circuit of 1 kilometer in the early morning calm of January 13, 1908, to the cheers of a small group of bowler-hatted friends and officials from the Aéro Club de France. At a height of some 10 meters and a speed of 60 km/hr, Farman cruised around a flagstaff placed at a distance of 500 meters from the starting point and, carefully bringing his machine about by means of its small rudder enclosed in the box-kite tail unit, returned in an uninterrupted flight to his starting point. To his personal profit—the 50,000-franc prize—and to the satisfaction of all those Europeans who doubted that the Wrights had flown, he had proved that controlled flight was feasible. In token of his achievement a true replica of Farman's biplane is preserved in the Musée de l'Air. A monument commemorating the event was created by the sculptor Paul Landowski.

Delagrange, less practiced in mechanics, did not make a sustained flight until March 14, 1908. While he ran Farman a close second, it was Farman who became the aviator of the year in French eyes. "What George Stephenson did for the locomotive," asserted one of his admirers, "Farman has done for the aeroplane." On July 6, six months after winning the Archdeacon–de la Meurthe prize, Farman pocketed the 10,000-franc Jules Armengaud prize for a quarter-hour flight—again at Issy. On the strength of this fresh attainment he issued a public challenge to the Wrights for a contest of speed and distance. Emboldened by his success, he was even willing to post a purse of $5,000—later raised to $10,000. Fortunately for Farman the Wrights did not accept, for it was their policy not to engage in publicity stunts; if they had, it is very likely that the big Voisin biplane would have been beaten. Instead Farman was persuaded to bring his ungainly machine to New York. Between July 31 and August 8, 1908, he made a series of exhibition flights at the Brighton Beach racetrack. Farman requested a runway one mile long and ½ mile wide—more than adequate for a modern jetliner's takeoff. But the best he could do the first day was a flight of 420 feet in 11 seconds at a height of 3 feet—and his remaining demonstrations were not much better. The American trip was an embarrassing mistake; neither the exhibitions nor the receipts came up to expectations, and Farman returned to France a sadder but wiser aviator.

On October 30, however, Farman regained his prestige by a daring cross-country excursion—the first ever of its kind—from the town of Bouy, near Châlons, to the cavalry grounds at Rheims. This overland journey of 27 km was accomplished in 20 minutes at an altitude of 30 meters, over telegraph lines, railway tracks, poplar groves, vineyards, and part of the village of Mourmelon. It led Farman to observe succinctly: "Theories about flying are, of course, very interesting; but what teaches is

Henry Farman making the first closed-circuit flight in Europe, Issy-les-Moulineaux, January 1909.

experience." Rheims never forgot the event. More than fifty years later, on June 6, 1960, a monument was erected on the spot where Farman landed: a marble shaft flanked by two columns, the left inscribed "Bouy" and the right "Rheims," with a facsimile of Farman's biplane carved on the surface.

On the basis of this instructive experience (which had been preceded by two successful flights, on September 29 and October 2, over the great open plain near Châlons), Farman felt qualified to start building his own biplane. At his Châlons workshop he did away with the fixed upright panels the Voisins had used for lateral stability—which not only tended to slow up the machine but caused it to drift sideways in a crosswind—and substituted four large hinged "ailerons," one each hinged to the rear spans of the upper left, lower left, upper right, and lower right wing panels. These drooped downward till the machine was in motion, when the airflow caused them to extend. This device, designed along the lines suggested by Alexander Graham Bell as a means of attaining lateral control, constituted a practical alternative to the wing-warping principle of the Wrights.

Farman also devised a light four-wheeled undercarriage with wooden skids to bring the plane to a stop, instead of the heavy metal chassis of the Voisin model; he lightened the tail and adapted the concept of a single forward elevator on an outrigger. At first fitted with a four-cylinder, water-cooled Vivinus engine, which was little more than an ordinary automobile motor pared down for the sake of minimal weight, the craft was later equipped with what was to become the most generally utilized of all air-cooled rotary motors—the Gnôme.

The robust Henry Farman pusher biplane was a success from its initial flight in April 1909, demonstrating qualities far superior to those of its Voisin predecessors. That it became the most popular biplane of the day in Europe takes on irony when it is recalled that Farman's decision to go it alone was the outgrowth of a dispute with Gabriel Voisin.

Farman, it appears had ordered and paid for a new Voisin machine; the builder, however, blandly proceeded to sell it to another customer— J. T. C. Moore-Brabazon (later Lord Brabazon of Tara) in England. It was Farman's fury at what he regarded as a double cross that caused him to design his own biplane. In a letter to Moore-Brabazon dated February 18, 1908 (found among the Englishman's papers, after his death in 1964, by C. H. Gibbs-Smith of the London Science Museum), Farman wrote of his "stupefaction" at this deal and of his resolve to have nothing more to do with the Voisins. From then on, his own success was constant.

Gabriel Voisin's image, on the other hand, was not improved by a book he wrote in 1926, reflecting his quarrel with Farman, in which he complained—in sorrow not unmixed with anger—that the plane which won the 50,000-franc prize in 1908 and which was exhibited at the first French Salon d'Aviation in 1909 bore the name of Farman, the pilot, with the name of Voisin, the maker, obliterated. More specifically he charged that a diorama in the 1926 air show illustrating the progress of powered flight had pictured the first three milestones as, successively, a kite, the Wright aeroplane, and the machine in which Farman accomplished his closed-circuit flight—with no credit given to Voisin. A distinct anti-American bias also was discernible in his attitude. Voisin always maintained that French aviation owed nothing to the Wright brothers; that it began on the sole foundation of its own resources; and that nobody had ever seen the Wrights fly before 1908. Nor was Voisin's personal reputation enhanced by a later publication, *Men, Women and 10,000 Kites*—a story of his life, which, an unkind critic has suggested, seems to have contained fewer kites than women. Voisin celebrated his eighty-eighth birthday in 1968, as testy as ever about the relative status of France and the United States in the procreation of mechanical flight.

At the same time that Henry Farman was developing his successful machine, his brother Maurice, a large, likable, cheerful man, set out on his own to create a biplane of the same type as Henry's—a type that because of the numerous struts and cross wires in the framework became generically known as the *cage à poules,* or "chicken coop." The M. Farman craft resembled the H. Farman, except that it carried the forward elevator on two protectively turned wooden skids that rose from twin sets of wheels, while the pilot (in contrast to the operator of Henry's plane) had a modicum of body protection in a short, fabric-covered cockpit, or "nacelle," with elementary windshield. Maurice was the only constructor who consistently favored the Renault water-cooled engine; and the standard Renault-equipped Maurice Farman machine was to prove a winner in several subsequent competitions. The protective nacelle, however, did not provide complete safety from the rear; for in pusher planes of this kind the engine and gasoline tank nestled in the small of the pilot's back and could prove a literally crushing weight in the event of a crash. In 1912 Maurice and Henry combined resources but continued the separate production of their distinctive machines, which for the year 1913 were estimated to total at least 300.

Maurice, the last survivor of the three Farman brothers, was an active pilot for nearly half a century. At eighty he was still flying regularly between his factory at Toussus-le-Noble (near the pioneer aerodrome of Buc, two miles southwest of Versailles) and his home at Barbizon, where he was in the habit of dropping in from the air for afternoon tea. Throughout his long and active life Maurice Farman never obtained a pilot's license, stubbornly declining to submit to a medical checkup. Since he had no right to fly alone, he was always accompanied by a trusted—and fully licensed—pilot. His machines gained a remarkable reputation for long-distance flight as well as for safety, and were a favorite of military authorities. Maurice himself eschewed the limelight, concentrated his career on the improvement of his planes, and carried his extreme modesty into death (at the age of eighty-seven) with the request that his funeral go unreported.

The Farmans and the Voisins did not alone hold the spotlight of aerial experimentation in France. In the early 1900's Louis Blériot, a well-to-do manufacturer of acetylene headlamps, had turned from automotive to aeronautical activity and begun a series of experiments that were to culminate in production of the most famous monoplane of its time. Inventor, constructor, and pilot, Blériot possessed both perseverance and an unquenchable ambition—a combination that helped him to overcome setbacks that would have utterly discouraged a less determined man.

Blériot was born in Cambrai on July 1, 1872. The patriarch of a family of six children, he was a stocky, dark man with a walrus moustache, brooding eyes, and an aquiline nose that gave him the look of a hawk. At considerable personal expense (he could well afford to spend some of the 60,000 francs a year he made from the sale of his headlamps) Blériot in 1901 attempted to put his ideas into practice with an ornithopter, one of the earliest flapping-wing machines. This optimistic effort to imitate the birds, like all such contraptions both before and since, was a total failure; so were his next two machines, built along the same complicated and impractical lines.

But in 1904 Blériot turned to the study of gliders and commissioned the ubiquitous Gabriel Voisin to build one for him. A partnership formed by the two men lasted through 1905 and 1906. The cellular box kite as developed by Hargrave—which also served as inspiration for Alexander Graham Bell—provided the principle of fixed vertical interplane surfaces in wing and tail units to give directional stability. After a preliminary experimental model, a biplane was built whose lower wing was foreshortened and connected with the upper surface by panels on either side; it had a single large tail unit. Towed behind a motorboat on the Seine, the Blériot-Voisin II ended its career in a sudden nose dive which all but cost Voisin his life by drowning (and which was recorded for posterity in a rare Gaumont newsreel). The Blériot-Voisin III was an even more curious contraption: the wings and tail unit were elliptical cells; two 25-hp Antoinette motors driving tractor propellers supplied the power; and the rudder was placed between the rear wings. Both Model III and its successor, Model IV—which had quadrangular cells for the forward wings, elliptical for the rear—were fitted first with floats, then with wheels; both were hopelessly ineffectual, whether tested from the calm waters of Lake Enghien, near Paris, or at Issy-les-Moulineaux.

Differences of viewpoint in both engineering and aerodynamic matters led to a parting of the ways between two obstinate and strong-willed theoreticians. Voisin continued on the same path with his biplanes, while Blériot branched into the investigation of monoplanes. In the spring of 1907 Blériot told his wife, the former Alice Vedère (a devoted companion and sympathetic audience of one), that he had great confidence in his newly designed Model V, a *canard*. One of the very earliest monoplanes, Model V had no tail; an elevator was located in front, and a propeller at the rear. It had paper-covered wings and a partially papered fuselage, and—borrowing from the Blériot-Voisin series—was mounted on two bicycle wheels and powered by a 24-hp Antoinette motor.

A Blériot canard, *one of the earliest monoplanes, at the Bagatelle cavalry grounds, 1907.*

Blériot's defection from the biplane school provoked a barrage of arguments pro and con. Proponents of the biplane looked askance at the relatively untried monoplanes, claiming that despite their simpler construction, single-surface machines would sacrifice too much lift. But before a valid judgment could be made, the *canard* lay a wreck. The same fate overtook Model VI—the overgrown Libellule, or "Dragonfly"—a tandem affair with wings covered by a tough parchment, whose outer tips could be pivoted for lateral control. For his Model VII, Blériot designed the now familiar cruciform type of the modern monoplane: engine and propeller in front of the wings, fuselage extending to elevators and rudder in the rear, and two-wheeled landing gear with shock-absorbing mechanism. But the controversy over the advantages and disadvantages of biplanes, monoplanes, even triplanes, was to continue for years.

Among the many problems confronting pioneers like Blériot was that of the air propeller. Data on the subject was meager. While some research had been done on airscrews for dirigible balloons (as derived from propeller-driven ships), their application to heavier-than-air craft was still in the experimental stage. Should the propeller be of small diameter and revolve at high speed, or on the contrary, should it be very large and turn relatively slowly? Should it be of wood, of metal, of metal-tipped wood? What should be its pitch—or angle of attacking the air? Should there be two, three, or four blades? What should be the weight, in relation to the weight of the motor? Should there be two propellers, revolving in opposite directions—thereby neutralizing the gyroscopic effect characteristic of one, but creating a hazard in case one propeller stopped? Should propellers be tractors or pushers? These were but a few of the questions that could be answered only by practical tests and experiments. From such empirical research it was found, for example, that in forming the blades of an air propeller it was best to curve them laterally. This could be done only by laminating, or gluing together, several sections of wood into one solid piece, allowing it to dry under heavy pressure, and then perfecting the surface finish.

Like so many others, Blériot made Issy-les-Moulineaux the main base of his flying endeavors. In the smoky bar of the Café des Sports, impressions were exchanged, plans made, tricks of the trade discussed—while in the music halls of Paris, topical tunes about aviation were beginning to catch on: *"Dans le Biplan," "Le Soldat Aviateur," "Dans mon Aéroplane."* One cold, gray day when visibility was low and the café was crowded, a stranger approached Blériot with a piece of advice. Introducing himself as a sailmaker, he suggested that the fuselage, or body, of a plane be covered with cloth to allow the air to flow over it more smoothly. It was an idea that Blériot incorporated into his Model VII *bis;* and although he never made a practice of completely enclosing the fuselage, the results obtained caused him to abandon his earlier use of varnished or unvarnished paper. Other constructors were to use treated, or "doped," fabric as a common technique, whether for wing or for fuselage covering.

Blériot was encouraged by the performance of his Model VIII, equipped with an eight-cylinder Antoinette motor and four-bladed flexible propeller. As in the case of everyone who aspired to fly, he spent much time rolling along the ground from one end of the field to the other. To get the feel of the machine and its controls, "grass cutting" was essential. Blériot soon graduated from feeble hops of a few meters to flights of 400 meters or more, at speeds estimated at nearly 50 m/hr. But while practicing a turn in February 1908, he allowed a wing tip to touch the ground, with the usual consequences: Model VIII was damaged beyond repair. Once more shrugging off the loss, Blériot wasted no time in building a new model—in the third version of which he flew without difficulty for 8 minutes 24 seconds at a height of 20 meters on July 6. On October 31 he achieved a short flight over the surrounding countryside (though with two unpremeditated stops due to magneto trouble).

But if he was gaining experience, Blériot was also acquiring a reputation for being accident prone. In crackup after crackup, however, he managed to escape more than superficial injuries. An awed British journalist proclaimed that "Monsieur Blériot is the most daring aviator in the world. . . . He has learned how to save himself when he falls, and when spectators rush up, expecting to find that he has broken a leg or an arm, he picks himself up and begins to give directions to his mechanics as to the necessary repairs. He seems to realize in an instant the exact extent of the damage and to know exactly what must be done to make the machine airworthy."

Blériot himself explained his luck this way: "A man who keeps his head in an aeroplane accident is not likely to come to much harm. What he must do is to think only of himself, and not of his machine; he must not try to save both. I always throw myself upon one of the wings of my machine when there is a mishap, and although this breaks the wing, it causes me to alight safely."

This preference for breaking a wing instead of a leg was miraculously demonstrated throughout Blériot's active flying career until, in December 1909, he met with one accident too many. While making an exhibition flight over Constantinople (now Istanbul), he fell onto a housetop and broke—in addition to both wings—several of his own ribs. Blériot took the incident as a stern warning, and flew no more after that. Instead he devoted his talents exclusively to the manufacture of the planes that bore his famous name.

Meanwhile still another design emerged from the workshop: Model IX. It was not a success, being overpowered with a sixteen-cylinder Antoinette motor of 65 hp and carrying a four-bladed propeller with radiators on either side of the fuselage. Model IX did, however, display a major contribution to aviation in the addition of the cloche to the control column. The cloche was a bell-shaped unit mounted on a universal joint, with wires leading to the stabilizing elements and elevator. Rudder movements, however, were controlled by the feet, resting on a bar. The cloche system was the original antecedent of the "joy stick," and as such the forerunner of all modern methods of control. It was patented by Blériot in April 1907 and replaced the various levers and wheels he had used before.

Then the improved Model X, which happily had completed a cross-country trip of 17½ miles in October 1908, was wrecked on November 4 when the unlucky pilot smashed into a tree in foggy weather. To Blériot's intense chagrin, other aviators—the Wrights, Gabriel Voisin, the Farmans, Delagrange—were all making news with their biplanes while he suffered one setback after another. Perhaps he had been wrong after all in his decision to stake everything on the monoplane. Blériot took stock of what he had to show for his efforts: he had developed a source of sufficient horsepower; he apparently had solved the problem of obtaining enough lift; and he thought he had mastered the related exigencies of stability, control, and ability to turn in the air. He had even perfected a swiveling undercarriage to allow the wheels of his monoplane to adjust to a bumpy pasture upon a rough takeoff or landing. What was wrong? He had spent a small fortune (as much as 780,000 francs, according to one estimate) developing an entire generation of different models. With a wife and children to maintain, he could no longer afford to borrow money for experiments ending one after another in costly crackups. As an alternative to giving up the game for good, there was one other course of endeavor: to win as many cash prizes as possible. In particular, Blériot set his sights on the £1000 offered by the London *Daily Mail* for the first flight across the English Channel.

At about this time two other adherents of the monoplane concept were trying to produce flyable machines: Robert Esnault-Pelterie and Léon Levavasseur.

Robert Esnault-Pelterie, an early Aéro Club enthusiast, was the son of a comfortably well-off cotton industrialist. Born in Paris on November 8, 1881, and educated at the Faculté des Sciences, he began his experiments with biplane gliders built using secondhand information of the Wright machines. However, since his data were incomplete, performance was faulty; and this led to erroneous conclusions about the Wright claims.

Esnault-Pelterie tried out his gliders on the beach and sand hills of Wissant, in the region of Calais; at one point he took the risk of being towed by an automobile, the better to study the mysteries of air pressure. His progress paralleled the advances of Blériot; by October 1907 he was flying his first monoplane—using internally braced wings, instead of a drag-producing system of external wires, and a lightweight engine of his own design. Known as the R.E.P. (after the inventor's initials), this was the first aeroplane with a completely enclosed fuselage of welded-steel tubing—a type well ahead of the times, embodying an engineer's idea of streamlining. Characteristically covered with red muslin, the R.E.P. ran along the ground on a single large bicycle wheel, with a smaller wheel at the end of each wing to maintain balance as the plane tipped to one side or the other, and a fourth wheel at the tail.

The designer was, like Delagrange, a man of many talents. He too was a sculptor—and also an engineer and a visionary who peered into the future like Jules Verne. Esnault-Pelterie predicted a rocket voyage to the moon that would take forty-nine days, and he wrote an incredibly far-sighted treatise on space travel. In addition to contributing a soundly designed monoplane to the early achievements of aviation, this creative bird-man devised, as early as 1906, an ingenious means of obtaining greater regularity of motor power—namely, the use of an odd number of cylinders. The original R.E.P. motor, with its fan-shaped "magic seven" cylinders delivering from 30 to 35 hp and weighing only 115 pounds, blazed the way for the development of other engines combining both power and lightness.

In an unpublished book of memoirs written for his son Michel, Robert Esnault-Pelterie tells of the crash that ended his career as a pilot. On June 18, 1908, he set out on a short trial flight; deciding to descend, he failed to realize that he should retard or cut the motor. The machine hit the ground at full speed. Despite an elastic seat belt the inventor was thrown against the fuel tank with such force that he broke one of its steel supports, while his right hip received a severe gash from another metal section. Suffering from shock and contusions, Esnault-Pelterie was found unconscious by a farmer who had witnessed the accident, and who revived the flyer with a stiff shot of cognac. From then on, the constructor left piloting to others; afflicted for years with the aftereffects of his injuries, which he feared might cause him to make some involuntary movement of the controls, he flew only as a passenger.

The R.E.P. monoplane, named after its designer, Robert Esnault-Pelterie, was the first plane with a completely enclosed fuselage. Constructed of welded-steel tubing, it was covered with red muslin. (Michel Esnault-Pelterie)

The R.E.P. in flight.
(*Michel Esnault-Pelterie*)

While the R.E.P. was the first flying machine employing a fuselage constructed of metal tubing, it was too heavy for the horsepower then available. It failed repeatedly to make good in flights at the Buc aerodrome, near Toussus-le-Noble, which the builder shared with his friend Maurice Farman after 1909. Not until necessary modifications had been made did the improved 60-hp R.E.P. begin to establish world speed records at the hands of the company test pilots, Ernest Laurens and Pierre-Marie Bournique, and win commendation for sheer excellence of construction.

The part played by Esnault-Pelterie in pushing the science of aviation several notches ahead is well illustrated by his patents of December 19, 1906; January 19, 1907; and January 22, 1907—which seem to have covered the combination of joy stick and rudder bar employed by Blériot. Technically, therefore, it might be claimed that Esnault-Pelterie was the originator of the cloche system. Because of the similarity in their thought along these lines, Esnault-Pelterie and Blériot agreed at one time to pool their patents. But after World War I the cloche principle was to entail almost as many complications among rival manufacturers as did the controversy between the Wrights and Curtiss over wing warping versus ailerons.

Esnault-Pelterie was also a pioneer with pneumatic shock absorbers, and he experimented extensively with steel, aluminum, and welded-steel tubing in the construction of both wings and fuselage. Financially, however, the R.E.P. was not within the range of an average purse—nor of an economy-minded military. In 1913 the workshop at Billancourt—one of the first fully equipped aircraft factories to be built in France—was sold to the Farman brothers, and one of the most forward-looking machines of the day passed out of existence.

If any one name was to signal the progress of European aviation more than another, it was that of a charming young girl—Antoinette Gastambide. First applied to a motor, then to a plane, her first name was immortalized by a happy set of circumstances. In the summer of 1902 Jules Gastambide, owner of an electric power plant in Algeria, was vacationing with his family at Etretat, an artist's retreat and resort town on the coast of Normandy, to which he had invited his friend the engineer and boat builder Léon Levavasseur for consultation. Levavasseur, corpulent and bearded, was the son of a naval officer. Born in Cherbourg in 1863, he had come to Paris at the age of seventeen to study at the Ecole des Beaux-Arts. His career had, however, been abruptly altered by the invention of the arc lamp. At the Ferranti-Patin factory where he had found work in that new field of illumination, Léon had launched also into an intensive study of gasoline engines.

On August 15 the four Gastambides and Levavasseur were taking a stroll along the cliffs overlooking the sea, watching the flight of cormorants. To Gastambide, Levavasseur remarked that man should be able to fly better than a bird—and that with Gastambide's financial help he would build an extralight engine specifically for that purpose, trying it out first in the popular new sport of motorboating. Then, turning to Gastambide's adolescent daughter, he added with a smile that the engine would be named "Antoinette" as a symbol of good luck. Thirteen days later a patent was applied for; and by 1904 Antoinette-powered speedboats were racing off with all the main prizes in Europe.

The natural outgrowth of such success on the water was adaptation of the motor for use in the air; and in May 1906 a commercial company was formed for the further production of Antoinette motors. Gastambide was president, Louis Blériot vice-president, and Levavasseur technical director. The V-8 water-cooled Antoinette of 24 or 50 hp, with a fair amount of aluminum in its construction for lightness, became the workhorse of the European pioneers—Ferber, Santos-Dumont, Delagrange, Voisin, Vuia, Blériot, Esnault-Pelterie, Henry Farman, and others. The engine had, however, an unfortunate tendency to quit working at critical moments. The Antoinette used a direct fuel injection system (of a type similar to that introduced by the Mercedes-Benz motorcar in 1955), but it was far from reliable. Fuel entered the eight cylinders through very small apertures; therefore the least impurity could cause a cylinder to clog, and constant inspection was necessary. Even when three filters were placed between the gasoline tank and the motor, Henry Farman found that a speck of dirt was sometimes responsible for an irregular performance. Nevertheless, despite difficulties with the carburetion and ignition, the Antoinette got many machines into the air; and if it did not always get them down as desired, it had no real rival until the advent of the radial Anzani and the rotary Gnôme in 1909.

Levavasseur never wavered in his belief that man should be able to fly better than a bird. In 1903, on an allotment of 20,000 francs from the special funds of the Ministry of War, he secretly constructed a large fly-ing machine of wood. With the aid of six associates the single-surfaced craft was transported under cover of night from Gastambide's shop at Puteaux to the spacious estate of some friends at Villatran, not far away. This rather weird bird, resembling a poised pterodactyl, had two propel-lers, one fore and one aft of the cockpit, turning in opposite directions; it was mounted on skids, with two small wheels designed to run along launching rails. Not even the presence of representatives from the Minis-try of War could induce it to fly, however; and the episode ended as clan-destinely as it had begun. The mortified inventor dismantled his appara-tus, removed the motor, and burned the remains. Subsequently a second model was constructed in the Lein shipyards on the banks of the Marne, but this was never tested.

A third machine, the Gastambide-Mengin—built at the request and at the expense of Jules Gastambide and L. Mengin, the two administrative heads of the company—bore trapezoidal wings; a long, plywood-covered, triangular-sectioned fuselage like a racing shell; a double rudder; and a four-wheeled undercarriage. On August 20, 1908, it became the first monoplane to make a two-man flight, piloted by R. Welferinger with Robert Gastambide (brother of Jules) as passenger. The following day it set another precedent as the first monoplane to complete a circuit. It was in this same general form that the Antoinette IV—the first of its kind to make a truly successful flight—appeared early in 1909.

When the company decided to expand its activities by the construction of aeroplanes as well as motors, it went against the advice of Blériot—with the result that he resigned. Blériot thought that the additional enter-prise would detract from sales of the engine; and while he personally was prepared to purchase one of the company's power plants for his own ex-periments, he preferred not to preside over what was tantamount to com-petition with himself in the manufacture of monoplanes. He may have felt justified when the 1907 model, despite its smooth lines and finely engi-neered construction, made an unpropitious showing, capsizing in a trial at Bagatelle in February 1908. As the year wore on, however, the Antoinette gave more promise with a series of short bounds at Issy-les-Moulineaux; and the 1909 model, although tricky and full of "temperament," per-formed very well indeed. An example of this type of plane, with gleaming copper radiator tubes along the sides of the fuselage, still evokes admiring comment in the hall of the Musée de l'Air at Meudon. Levavasseur al-ways kept his touch, and no model ever rivaled the Antoinette in grace and beauty of design.

As presiding genius of one of the greatest aeroplane firms of the day, Levavasseur was among the first to make money out of aviation. But he received only a taste of success. Perhaps overly confident of his own bril-liance, he did not always take kindly to the advice of others and was reluc-tant to acknowledge the possibility that he himself might ever be wrong. While the years immediately ahead were to bring his monoplane renown, an advanced cantilever low-wing model that he built for the French military trials of 1911 was a failure. Although it embodied radical ideas of streamlining (including streamlined landing gear, or "pants"), and allowed access to the engine in flight, it was insufficiently powered by a 90-hp motor; it was also too hastily produced, being technically at fault in several respects. Soon thereafter the company's finances deteriorated, and the Antoinette was swallowed in the maw of bankruptcy. In 1921, after the lessons of World War I, its inventor was to produce one more machine—a monoplane with variable wing surfaces that was too far ahead of the times to impress potential customers. Levavasseur died at the age of fifty-nine, in greatly reduced circumstances, on February 24, 1922.

The European aviation scene in the spring of 1908 was completely dominated by the friendly rivalry of Henry Farman and Delagrange. Farman had gone to England in February, seeking a suitable flying ground to demonstrate his Voisin machine, but had found none to his liking and returned to continue operations on the familiar territory of Issy-les-Moulineaux. Delagrange was demonstrating in Italy. On May 15, in response to a special invitation, he gave that country its first glimpse of a flying machine, exhibiting his Voisin first at Rome and a few weeks later at Milan and Turin. At Turin on July 8 he took up Thérèse Peltier, his sculptress pupil at the Beaux-Arts, for a ride of about 200 meters at a height of some 4 meters above the earth. Thérèse thus became the first woman in the world to ascend in an aeroplane; she went on to make a few flights alone but eventually abandoned the sport. While these exhibitions were not a financial success, they led to the formation of the Società Aviazione Torino, as well as similar embryonic flying clubs in Rome, Milan, Naples, Florence, Ferrara, Palermo, and Verona.

As summer approached, the names of Farman and Delagrange appeared with increasing frequency in the newspapers. All tentative attempts to emulate the birds were faithfully chronicled. At Issy as elsewhere, there were fledgling experiments by other constructors—such as Alfred de Pischoff, an Austrian resident of Paris who had first flown his tractor biplane on December 5 and 6, 1907. But Europe was to see some real flying before long.

Rumor now had it that the Wrights were coming abroad, and that on their arrival some aerial feats would take place far surpassing any previous exploits. Rebuffed at home, the brothers were indeed making plans to demonstrate their machine in France. Wilbur arrived in May 1908 and picked up the biplane which for a year had lain idle in its crate at customs in Le Havre. Through the courtesy of Léon Bollée, a jolly, goateed manufacturer of automobiles, arrangements were made to use the race-course at Hunaudières—a sandy, open terrain surrounded by pine trees near the quiet town of Le Mans, about 130 miles west of Paris. Le

Mans—today the annual scene of a grinding twenty-four-hour auto race—was the site of Bollée's factory, and arrangements were made to assemble the plane in a corner of his establishment. Weeks of tinkering went by as Wilbur methodically prepared for his first exhibition on foreign soil. After five years of public indifference, skepticism, and even hostility at home, Wilbur seemed unconcerned about the suspense building up in France over the impending test. He guarded his invention behind closed doors, as if surrounded by spies, and lived, ate, and slept next to the machine. Conscious or not of the significance of the moment, the American inventor-pilot was on the point of recognition as well merited as it was long overdue.

On Saturday, August 8, 1908, the impatiently awaited trial took place before a weekend throng of spectators—many surveying the scene from the vantage point of trees, some with picnic lunches, and all keyed up in the belief at last that something momentous was going on in the makeshift hangar at the edge of the track. As reported in the Paris *Herald,* the plane looked frail in relation to its bulky motor, which balked at starting on the first two or three tries. A tear on the lower wing had been noticeably repaired with a patch and glue. Wilbur Wright himself, far from dramatizing the occasion, was dressed as if for a stroll: gray suit; high, starched collar; and a cap. When the weight on the catapult dropped, the plane lunged forward on its monorail "like an arrow from a cross-bow shot into the air. . . ." Within fifty feet of the start, recounted the *Herald* next day, "the machine rose to a height of eight to ten meters, circled twice, took turns with ease at almost terrifying angles and alighted like a bird. The flying time was 1 minute, 45 seconds."

It was a stupendous vindication. No longer was the term *bluffeur* to be heard. As the voluble balloonist Surcouf put it to the members of the Aéro Club de France: *"C'est le plus grand erreur du siècle!"* ("It is the greatest error of the century!") Disbelief in the Wright claims had indeed been a colossal mistake, and the French were magnanimous in admitting it. "This is the beginning of a new phase of mechanical flight!" exclaimed Blériot. "Wright is a genius. He is the master of us all." The newspaper *Figaro* said: "It was not merely a success but a triumph, a conclusive trial and a decisive victory for aviation, the news of which will revolutionize scientific circles throughout the world." When Wilbur made flights of better than 3, 6, and 8 minutes' duration in the next few days, the while executing circles and figure eights, Delagrange threw up his hands and exclaimed theatrically: "We are beaten! We just don't exist!" "That Wright is in possession of a power which controls the fate of nations," said Major B. F. S. Baden-Powell, secretary of the British (later Royal) Aeronautical Society, "is beyond dispute."

Tributes poured in on every side. France was flooded with picture postcards portraying Wright and his aeroplane—the start of a souvenir card craze that was soon to provide the public with a likeness of every plane and pilot in the country. When the French ambassador to the United

States arrived in New York, he announced that the modest Mr. Wright was the greatest hero to be found in France. On July 17, 1920, after the dust of World War I had settled, a monument honoring Wilbur Wright and the "Precursors of Aviation" was unveiled in the historic Place des Jacobins at Le Mans. The monument, the work of the French sculptor Paul Landowski, depicts the Greek myth of Daedalus and Icarus.

But confident as he was in the future of flight, Wilbur was not wholly accurate in his predictions off the cuff. He thought that ultimately man would fly without motor power, "like a buzzard"; and that the biplane—not the monoplane—was the answer to all aeronautical problems because of its greater strength in construction and additional wing surface area.

After Wright's initial success at Hunaudières, the military authorities in Paris let it be known that they would be honored if the flights were transferred to the open spaces of the artillery testing grounds at Auvours, ten miles away. The invitation was accepted. Here, on September 16, Wilbur took up a passenger—the young French balloonist and aeronautical engineer Ernest Zens. Zens, with his brother Paul, was soon flying a machine of his own design, powered by a 50-hp Antoinette motor. Wilbur's flights in the region are commemorated by a plain monument that stands at one corner of the Auvours proving grounds.

President Theodore Roosevelt's interest in the Wright claims had led at last to the delivery, on August 20, 1908, of a "heavier-than-air Flying Machine," in accordance with the Army's contract specification (No. 486). While Wilbur was demonstrating in Europe, Orville brought the 1908 Flyer to the drill field at Fort Myer, Virginia; there, on September 9, he had made a flight of 1 hour 2 minutes 15 seconds—the first time that an aeroplane had flown for more than an hour. Later the same day he carried aloft an Army balloonist, Lieutenant Frank Purdy Lahm, who thereby became the first officer to fly as a passenger in an aeroplane. The flight of 6 minutes 24 seconds established a world's record for two men. After this, Orville made flights almost daily; and as the word spread, people flocked to Fort Myer by the thousands.

Tragedy, however, stopped the demonstrations. During a flight on September 17, the right propeller fouled a guy wire and broke off, causing the machine to crash from a height of 75 feet. Orville Wright and his passenger, twenty-six-year-old Lieutenant Thomas Selfridge, were taken to the Fort Myer hospital. Orville was seriously injured; but Selfridge was so grievously hurt that he died that evening—the first person in the world to be killed in an aeroplane accident. Thus the completion of the Army tests had to wait until the following year. If Selfridge had lived, his experience as an engineer and his background as a graduate of West Point, not to speak of his devotion to flying, would have served his country well. In 1965 his name was inscribed in the National Aviation Hall of Fame at Dayton.

Progress during the month of September 1908 was on the whole ex-

While his brother demonstrated planes in Europe, Orville Wright brought his 1908 Flyer to Fort Myer, Virginia, to conduct a series of trials for the Army. On September 9 he made a flight of 1 hour 2 minutes 15 seconds— the first time an aeroplane had remained aloft for more than an hour. (National Archives)

The Army tests were discontinued after a fatal flight on September 17. Orville Wright, at the controls, was seriously injured but recovered. His passenger, Lieutenant Thomas Selfridge, died that evening—the first person in history to be killed in an aeroplane accident. (Air Force Museum)

*The wreckage of the
plane carrying Wright
and Selfridge, a few
minutes after the crash.
(Air Force Museum)*

traordinary, considering that the year had opened with a circular flight of
only 1 kilometer and of little more than a minute in duration—and that
only four men were involved at the time: Wilbur Wright, Henry Farman,
and Delagrange in France; and Orville Wright in America. The length of
time a man could stay up was increasing rapidly, as shown by a compilation of the official records for the principal flights during that month:

Date (*Sept. 1908*)	*Aviator*	*Time* hr	min	sec
5	Léon Delagrange	0	29	53⅘
5	Wilbur Wright	0	36	14⅗
6	Léon Delagrange	0	28	0
9	Orville Wright	0	57	31
9	Orville Wright	1	2	30
10	Wilbur Wright	0	21	43⅖
10	Orville Wright	1	5	52
11	Orville Wright	1	10	0
12	Orville Wright	1	14	20
16	Wilbur Wright	0	39	19
21	Wilbur Wright	1	31	25⅘
24	Wilbur Wright	0	54	3⅕
28	Wilbur Wright	1	7	24
29	Henry Farman	0	43	0
30	Henry Farman	0	35	36

Later Wilbur began taking up others—including his first woman passenger, Mrs. Hart O. Berg. She was carried aloft for 2 minutes 3 seconds on October 7, and thereby started a fashion for dressmakers. To keep her skirt from blowing about, Mrs. Berg tied a rope just above her ankles; the difficulty she experienced in walking after the landing was observed by a Paris couturiere—and thus the "hobble skirt," with its many variations, was born. Another pioneer passenger was George Dickin of the Paris *Herald,* the first newspaperman to fly. A year later Wilbur was also to take up the first woman passenger in the United States—Mrs. Ralph Van Deman, an attractive friend of his sister Katharine—who flew with him at College Park, Maryland, on October 27, 1909.

For the first time, too, Wilbur began giving lessons, as part of a contract the Wrights had concluded with Lazare Weiller, a wealthy Frenchman, for the formation of a syndicate to manufacture the Wright plane under license in France. The French company was named the Compagnie Générale de Navigaton Aérienne, and its sales outlet was known as the Société Ariel. The first three students were Count Charles de Lambert, Paul Tissandier, and a forty-five-year-old army officer, Captain Lucas Girardville. Before transferring his "school" to the milder climate of Pau, a resort town in the Pyrenees (where a second monument by Landowski was to be dedicated in his honor in 1932), Wright went after a number of cash prizes. On November 13 he won 1000 francs from the Aéro Club de la Sarthe for an altitude record of 70 meters; on December 18 he took another of the club's prizes by reaching a height of 100 meters—a new world's record. On December 31, the last day he flew at Le Mans, Wilbur won a cash prize of $3000, donated by the firm of Michelin Brothers. The Michelin competition (which of course served well to advertise the rubber tires by that name) was set up as an annual affair over a period of ten years for the flight of longest duration, provided that each year's flight was at least double the previous record. Wilbur was an easy winner of the first contest, staying up for 2 hours 20 minutes 23.2 seconds. Like so many other sporting events, the Michelin contest was to be overtaken by World War I, the last winner being Emmanuel Hélen on October 22, 1913.

The winter season of 1909 at Pau proved an outstanding success for the Wright brothers. Wilbur was joined by Orville and Katharine. Among the notables who displayed intense interest in the embryonic school of aviation were King Alfonso XIII of Spain, King Edward VII of England, and former British Prime Minister Arthur Balfour. Captain Alfred Hildebrandt, representing the owner and publisher of the *Berliner Lokal Anzeiger,* paid a visit to the brothers' camp and arranged for a demonstration in the German capital later in the year. The Wrights were to be given a substantial fee, and the general public was to be invited as the newspaper's guests; both the paper and the Wright biplane would benefit from the publicity.

In April 1909 the Wrights moved to Rome with one of their machines

Largely unrecognized in their own country, the Wright brothers were widely hailed on their trip to France in 1908. The following spring their hometown of Dayton, Ohio, honored them with an official celebration. (National Archives)

and thereupon showed Romans what it really was like to fly. At nearby Centocelle (to be established the following year as a civil and military aviation center) Wilbur took up the American ambassador, Lloyd C. Griscom; fascinated King Victor Emmanuel III with the gyrations of his plane; and, as part of his contract for a Wright-licensed company in Italy, provided instruction for two neophyte pilots. These were a young Italian naval lieutenant named Mario Calderara (later air attaché at the Italian Embassy in Washington) and Lieutenant T. U. Savoia of the Italian Military Engineers, who in June 1910 became the first to fly over Rome—from Centocelle to Lake Bracciano—and whose name would one day become famous in connection with the important manufacturing firm of Savoia-Marchetti.

The success of the Wrights in Europe had finally induced America to give the fathers of flight their due, when, fresh from their foreign conquests, the brothers returned to Dayton. "Guns Boom, Whistles Blow, Crowd Cheers," the evening *Herald* reported on May 13, 1909. "Amid the roar of cannon and screeching of whistles that made up the demonstration that men, women, boys and girls tendered in honor of their wondrous exploits with the aeroplane abroad, Wilbur and Orville Wright and their sister Katharine came home today." The entire front page was given over to the town's two distinguished citizens, with pictures whose caption ranked them with Fulton, Edison, and Morse. On June 17 and 18, a delirious official welcome took place, with bands playing, bells ringing, and a lavish display of fireworks. Praise knew no bounds. From the stone-hearted apathy of the years just past, the public in a surge of patriotism went to the opposite extreme.

As soon as the celebration was over, the Wrights left for Washington to complete the series of trials for the U.S. Army that had been interrupted the year before by the disastrous crash of the Flyer. The Wrights' contract with the Government provided that the machine must do at least 40 m/hr. When the trials were resumed in 1909, Lieutenant Lahm was again a passenger, in a new and improved Flyer piloted by Orville. Together, on July 27, they set a world record for pilot and passenger of 1 hour 12 minutes 40 seconds. Three days later, accompanied by Lieutenant Benjamin Delahauf Foulois, Orville made the first cross-country flight in America: from Fort Myer to Alexandria and back, a round trip of 10 miles at the fast average speed of 42.58 m/hr. Following these official tests, the 1909 Flyer was accepted and purchased by the War Department for $25,000, plus a bonus of $5,000 for exceeding the stipulated speed. It thus became the world's first successful military aeroplane. As part of the sale contract, Wilbur trained two Signal Corps lieutenants, Lahm and Frederic E. Humphreys, to operate the Flyer. (They received their instruction at College Park, Maryland, between October 8 and 26, 1909.)

Foulois, later a brigadier general, recalled that when he was ordered to Fort Sam Houston, Texas, with Signal Corps aeroplane No. 1 to teach himself to fly, he was admonished to take along "plenty of spare parts." On that machine he had his first solo flight, first takeoff, first landing, and first crackup—thereafter receiving instructions from the Wright brothers by mail whenever he needed advice on some as yet obscure aspect of becoming a pilot. This premier military plane remained in active service until 1911; it now shares honors with other early aircraft in the National Air and Space Museum in Washington. A plaque on the parade ground at Fort Myer commemorates the 1909 trials.

The German obligation was fulfilled by Orville Wright in August and September of 1909 at the Tempelhof field, then a military parade ground on the outskirts of Berlin. He was not, however, the first to fly at Berlin: Armand Zipfel, a twenty-six-year-old Frenchman from Lyons, had made the first ascent, under auspices of the *Lokal Anzeiger,* in a Voisin on November 24, 1908. And the German flyer Hans Grade had succeeded in making a hop at Magdeburg on November 2, 1908, in a tiny triplane. But nothing like the grace and ease of Orville's flying had been seen in Germany before. That country had hitched its wagon to the star of Count Zeppelin's ponderous rigid dirigibles, and competition in the heavier-than-air class was lacking.

Alone and with Captain Paul Engelhard of the German army as passenger, Orville broke records and won the adulation of enormous crowds. He flew for Kaiser Wilhelm II and even gave a ride to the crown prince—the first member of a royal family to fly. He also concluded a contract for a German Wright company, and as a condition thereof trained Engelhard and the civilian Fridolin Keidel as flyers at the Bornstedt field, near Potsdam. Engelhard was the chief aviator of the German Wright firm until he was killed in a crash at the Johannisthal aerodrome on September 29, 1911. The machine that Orville used in his Berlin demonstrations, built in Dayton and assembled in Germany, is housed today in the Deutsches Museum at Munich—the only surviving example of the standard Wright Model A biplane of the years 1907–1909.

CHAPTER FOUR

There Are No Islands Anymore

If, in the saga of powered flight, a moment can be fixed representing the turning point when doubt as to the practicality of the aeroplane gave way to excited acclaim, it must be a six-week period in the summer of 1909. The aviators of that year unexpectedly became instant heroes—sharing the limelight with Robert E. Peary, who had discovered the North Pole on April 6. Three great events of 1909 proved conclusively that man could fly: the English Channel was crossed by air; the first international air meet took place in France; and the first race was held for the Gordon Bennett trophy, the blue ribbon classic of early aviation.

When the London *Daily Mail,* on an inside page of its issue of October 5, 1908, casually offered £500 for the first flyer to cross the English Channel in either direction, few took the announcement seriously. After all, it was only nine months since Henry Farman had completed his 1-kilometer closed circuit—an astonishing achievement at the time. Of course, the "ditch" had been crossed before by air. Beginning in 1785 with the Frenchman Jean-Pierre Blanchard and his American backer, Dr. John Jeffries, who bridged the gap between Calais and Dover in 2 hours 20 minutes, it had been a favorite project for free-drifting balloonists. No fewer than thirty-six such traverses, in one direction or the other, were recorded over the next 124 years by lighter-than-air craft.

The year 1908 ended, and with it the offer—without takers; but the prize was reestablished in 1909 at double the original figure. Among other provisions, it was specified that the flight would have to be made between sunrise and sunset; that no part of the machine should touch the sea during the crossing; that the machine should be heavier than air, and without the assistance of a gasbag or similar contrivance; that at least forty-eight hours' advance notice of an attempt should be given to the *Daily Mail;* that a contestant must furnish proof of having successfully flown before; and that the contestant must renounce any claims for damage to machine or person in case of accident.

Such an opportunity could not long be ignored by the early birds, who were determined to prove their faith in flying by deeds no less than words.

One of them was Hubert Latham, a twenty-six-year-old Frenchman of English descent, who—partly because of his parents' ties with Britain—resolved early in June to make the Channel crossing his personal affair. Santos-Dumont was another; but the Brazilian withdrew when friends insisted that the diminutive Demoiselle was incapable of such a madcap venture. An Englishman named Seymour was reported to have readied a hangar and takeoff space for his Wright biplane at Le Touquet; and at Boulogne, Charles Stewart Rolls (cofounder of the Rolls-Royce motor firm), holder of British license No. 2, also hoped to gain the honor for England with a Wright biplane. Serge de Bolotoff, of Bulgarian nationality (known also in London as a Russian prince), who had thrown his hat into the ring as early as November 1908, contemplated a crossing from the port of Saint-Lô with a triplane ordered from Voisin Frères especially for the purpose. Powered by a 100-hp Panhard-Levassor engine, the whole apparatus weighed more than a ton. However, when the facts of flight impinged on the fancy, Bolotoff was moved to scratch his name from the list.

The most serious contender in addition to Latham appeared to be Count Charles de Lambert, a romantic-looking Russian of French ancestry whose father and grandfather had served as generals in the army of the czar. Born on the island of Madeira in 1865, de Lambert was a large man with a bountiful reddish moustache and steel-blue eyes. A sportsman and inventor who had engaged himself in studying the giant steam-propelled flying machine devised by Hiram Maxim, he was the first European to receive a pilot's certificate under the tutelage of Wilbur Wright. At Wissant, de Lambert assembled two Wright biplanes, using the 24-hp Bariquand et Marre motor, and prepared to take off at the first break in the weather. In his high celluloid collar, chauffeur's cap with flaps buttoned on top, and large mourning arm band worn for a member of his family, the count was the center of a respectful group of onlookers as he and his mechanics went about their tasks. All the count's preparations, however, went for naught: he damaged one machine in a trial and was overtaken by events before he could get the other in shape for the Channel crossing.

The first of the would-be Channel hoppers actually to get off the ground was Latham. Born in 1883, he was a cultured Parisian, fluent in English and German, who had taken a degree at Balliol College, Oxford. A bachelor sportsman who moved in aristocratic circles, Latham indulged a yen for adventure by hunting and exploring in Abyssinia. His paternal grandfather—the first Latham to settle in France—had come from Lincolnshire, England, having made a fortune speculating in indigo. The family seat was established at Maillebois, near Chartres; an imposing château in the Norman style, it was surrounded by an extensive park. On his mother's side a member of the Mallet banking family, Latham was related to the German chancellor, Bethmann-Hollweg.

"I am a man of the world," Hubert was to answer President Fallières of France when asked what his occupation was, apart from flying. In a full-length portrait hanging in the château living room today, he appears as a good-looking, rather pale and slim, debonair person—with fine, regular features, neck choked by the high, stiff collar of the time, and brandishing the long cigarette holder which was as much a part of him as the cigar was to be of Winston Churchill. Among the household treasures is the first duck ever shot from an aeroplane—a souvenir of a flight he made in California. Proudly displayed nearby is a handsome silver trophy awarded to Latham by a Berlin air club for the first overland flight in Germany, completed on September 27, 1909, between the embryonic flying fields of Tempelhof and Johannisthal. On a wide expanse of meadow adjoining the château is a stone marker to commemorate the day that Hubert first flew down from Paris for lunch, while a monument erected in the nearby village does him further homage.

Latham was as well known for his checked cap and wristwatch as for his cigarette holder. His insouciance, coupled with what seemed to be utter disregard for personal safety, foretold an early grave for him in the eyes of his fellow flyers—whose own life expectancies were certainly not high. Yet there was a reason for his bravado: Latham had tuberculosis, and himself felt that he had not long to live. He did indeed die accidentally before his time—but not in an aeroplane.

Over the short span of his active career Latham burnished the family name to its lasting pride. His connection with mechanical flight began in 1908, when he became enamoured of the Antoinette—that beautiful, aerodynamically advanced brainchild of Levavasseur, whom he met through his friend and neighbor Gastambide. In the years preceding, Latham had crossed the Channel as a passenger in a free balloon with his cousin, the aeronaut Jacques Faure, and had won races at Monaco in speedboats powered by Levavasseur. When Levavasseur introduced him to the Antoinette entrepreneurs, he was quick to see in their arresting machine the possibility of another exciting form of sport. On February 25, 1909, the partners approached the well-to-do Latham with a proposal that he become their chief pilot and hence serve as standard-bearer for the house of Antoinette. Ready as usual for adventure, Hubert needed no urging. After a number of semisuccessful trials under the instruction of Eugène Welferinger, head of the firm's bureau of studies, Latham succeeded in making a perfect flight of 10 minutes at Châlons-sur-Marne.

Progress thereafter was rapid. On June 5—novice no longer—Latham broke the world record for monoplanes, as well as the French record for mechanical flight, by staying aloft for 1 hour 7 minutes 37 seconds, until a rainstorm forced him down. Enlarging upon his success, the next day (June 6) he won the Ambroise Goupy prize by covering a straight-line course of 6 km in 4 minutes 13 seconds. On June 8 he demonstrated the novel feat of the volplane—a glide with motor cut off—and capped it by giving a ride to a representative of the *Daily Mail*. On June 12 he flew 40 km in 39 minutes before a parliamentary group for the advancement of aeronautics. On June 23 he took off in a breeze of 25 km/hr—a convincing demonstration of the Antoinette's natural stability. With this much—and no more—experience behind him, Latham was inspired to try for the *Daily Mail* prize for a flight across the Channel.

Without more ado the Antoinette IV was dismantled, crated, and sent by train to Calais; from there it was transported to the site Latham had selected as a jumping-off point for England. His base of operation was the village of Sangatte, near the abandoned headquarters of an enterprise that had foundered in 1883: Lord Grosvenor's Channel Tunnel Company. In that abortive attempt to link the shores of France and England, Latham saw a superb twentieth-century opportunity—a second effort, this time by air, which if successful from the French side would not only puncture British pride but strike a telling blow at British insularity. He made himself comfortable at the Grand Hôtel in Calais while Levavasseur, supervising a corps of ten mechanics, settled into the only inn at Sangatte. There an out-of-tune piano in the back room of the café helped to enliven a wait on the weather.

Fog, rain, wind—those archenemies of the pioneers—made each day more exasperating than the one before. Daybreak found Latham up, scrutinizing the sky; as the morning wore on, he would shuttle between Calais and Sangatte, checking every detail of his plans, greeting visitors and well-wishers, chain-smoking his cigarettes. Conditions, however, remained unfavorable for flight. As days went by and the delay continued, the aviator's nerves became visibly frayed. Never had there been such abominable weather.

A wireless hookup arranged by the *Daily Mail* facilitated the exchange of meteorological reports between Calais and Dover. Tension ran high on both sides of the Channel. While the elements conspired to increase the suspense, an expectant public was fed hourly bulletins: "Latham Awaiting His Chance"; "Latham's Disappointments"; "Latham's Hope—Very Little Wind." Dover was filled with the curious, scanning overcast skies. A flotilla composed of yachts, motorboats, and a portion of the British Home Fleet had assembled at the port; a White Star liner about to leave for America delayed its departure at the request of the passengers. A special flag was installed on the roof of the Lord Warden Hotel to signal the takeoff as soon as word was flashed. The whole world was watching and waiting.

In this ambience of hope, skepticism, and disappointment, Latham decided upon a trial flight. On July 13, in the calm before sunup, he took his Antoinette into the air for a few minutes. Returning to his starting point, however, he found the meadow overrun with spectators and was forced to land in a ploughed field, damaging the plane's undercarriage. The following day he discovered that his improvised workshop had been entered and the storage batteries stolen. Another delay ensued while they were replaced.

At last, early on July 19, the wind dropped abruptly. The morning mist gave way to present a visibility of 10 miles, and a wireless message from Dover confirmed that the outlook was excellent. Latham would lose no more time; this was the moment. At 6:42 A.M. a message crackled from Sangatte, advising England—and the world—that the great attempt had begun.

First circling the solitary chimney of the old Tunnel Company, Latham crossed the cliffs of Cap Blanc-Nez and headed northwest. By the time he reached an altitude of 1000 feet he had overtaken the Harpon—a destroyer escort provided by the French navy—which had put to sea at the first signal that the flight was on. Everything was functioning normally. For the first time in history a heavier-than-air machine had set out for a destination across an open body of water. Perhaps in exultation at the thought, Latham indulged his avowed weakness: a cigarette. As he later noted in a reminiscence of the flight, he was simply unable to abstain from what he called this "light stimulant." Some think that Latham's ever-present cigarette contained marijuana; if it did, it may have given him the composure and the nerve needed to accomplish all he did.

But for those who waited on the heights of Dover, the next few hours were filled with foreboding. Urgent queries were sent over to Sangatte: "Seven twenty-three. Anything yet?"; "Seven forty-six. Nothing in sight, requesting assistance, fishery cruiser to search"; "Eight six. Very anxious here. Cannot see torpedo boat [sic] or Latham. Can you?" At 8:11 Sangatte replied that the Antoinette had been lost to sight ten minutes after departure, and that a large crowd was waiting for news there too. Half an hour later, word spread that the plane had fallen into the sea. Not until 10:23 did the Dover authorities learn for certain that the pilot was safe.

What had happened? Probably a rupture of a fuel line, or failure of a spark plug; the uncertain engine of the Antoinette had just stopped. Seven miles out from the French coast, Latham had seen his propeller begin to turn more and more slowly. Still within sight of the Harpon, he had made another of his long, skillful volplanes; pancaked into the sea; and stoically awaited rescue. Perched on the slender fuselage of the floating plane, feet dangling in the water-filled cockpit, hand on the left control wheel to keep his balance, he had imperturbably lighted still another cigarette. Within twenty minutes the Harpon was alongside to take him off in a lifeboat; but hoisting aboard the monoplane completed the wreckage of the already damaged machine.

Latham, who had bet 17,000 francs that he would cross the Channel before August 1, immediately telegraphed the Antoinette factory at Puteaux to send him another plane. He had good reason for haste. Another strong contender—Louis Blériot—had entered the lists that very day, notifying the Daily Mail of his intention to try for the prize within the week.

Blériot's dramatic arrival on the scene resembled the unexpected appearance of Charles A. Lindbergh, nearly two decades later, in the race to be first across the Atlantic. The eleventh-hour financial support that enabled Blériot to cast his lot in the Channel competition came wholly by chance. Toward the end of June, obsessed by the need for finding more cash if he was to continue in aviation, Blériot had made up his mind to enter as many contests as possible in a final gamble to gather prize money. On July 1 his wife, Alice, was visiting friends when she providentially saved the life of a young boy, the son of a Haitian planter named Laraque, whom she caught teetering on the edge of a high balcony. That evening the Laraques called upon the Blériots to express their thanks; and during the course of the visit it developed that the Haitian had a deep interest in aeronautics. When he learned how hard pressed Blériot was—the motor for his Model XI had not yet been paid for—Laraque spontaneously wrote a check for 25,000 francs, asking only a receipt in return until some settlement could be worked out. Blériot, overjoyed, promised Laraque half the *Daily Mail* prize should he win it.

During the first week in July, Blériot took a number of lesser prizes in a small meet at Douai, including the Corderie prize for flying 1500 meters in a closed circuit. On July 3, in a flight of 47 minutes with his Model XII—designed with an aileron type of control instead of wing warping—he covered a good 42 km. This was a sufficient achievement, he thought, to warrant an attempt to cross the Channel at its narrowest point—only 38 km—at the Strait of Dover. On the July 3 occasion, however, the protective asbestos lining of the plane's exhaust pipe had torn loose, and Blériot's left foot had suffered a third-degree burn.

Despite the injury he managed on July 4 to win a further prize, offered by Madame Ernest Archdeacon—who was not to be outdone by her husband as a patron of aviation. This time he took his Model XI (controlled by wing warping) 24 times around the course in 50 minutes at a height of 40 meters. On July 13 at Etampes he won the Aéro Club de France Prix du Voyage of 4500 francs by flying the 41 km to Orléans in an elapsed time of 44 minutes 30 seconds, with one stop; and on July 18 he was back at Douai, again with Model XII, to win the Mathieu prize for speed—covering 2 km in 2 minutes 29 seconds. Unhappily the same mischance befell him as before: his left foot was severely burned—and before the previous injury had had time to heal. Since this machine had given him so much trouble, Blériot discarded it for the Channel attempt in favor of Model XI. After being shipped to Calais on a flatcar, wings folded against fuselage, the monoplane (weight 245 kilograms empty) was transported by cart to the chosen takeoff point—in a field at the nearby village of Les Baraques.

Model XI was the prototype of the most famous monoplane of the day. This highly successful design was soon to carry the name of Blériot to the distant corners of the earth. Although originally fitted with a 30-hp R.E.P. motor, in its final form it was powered by an engine featuring notable innovations. This three-cylinder, radial, air-cooled power plant was the invention of a former Italian bicycle racer and trainer living in

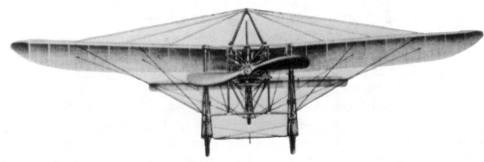

Sketch of Louis Blériot's model XI monoplane, the first flying machine to cross the English Channel. Blériot made the 22-mile trip from Calais to Dover in 38 minutes, on July 25, 1909.

Paris named Alessandro Anzani, who manufactured motorcycles equipped with a lightweight two-cylinder engine cast in the form of a "V." A tendency for his experimental three-cylinder model to overheat was conquered by the resourcefulness of Anzani's chief mechanic, Francesco Santarini; but the 25-hp Anzani became as notorious for its propensity to throw oil and turn the air blue with acrid fumes as its inventor was known for his racy language. For some time Santarini had been trying to find a better means for discharging the exhaust, to keep the engine from getting too hot and to improve its general efficiency. One night the mechanic disassembled the cylinders and drilled fourteen perforations of an 8-mm diameter in the cylinder walls, to allow an outlet for the expanded gases after combustion. The scheme worked, and the engine remained cool—but at the price of consuming an inordinate amount of oil, with copious amounts spilled in the process. One early Blériot-Anzani pilot coped with the "fallout" by attaching a wad of cotton to a string suspended from his goggles, with which he wiped the lenses clean from time to time.

The propeller had been selected with care, for by now it was realized how much depended on efficient use of the power transmitted by a motor. Wooden propellers were subject to the effects of climate; splitting of the blades while in motion could be the cause of a serious accident. For his Model XI, Blériot used a laminated-walnut tractor 6.6 feet in diameter, a fairly popular model built by the well-known firm of Chauvière.

Blériot was hardly in the best physical condition when, on crutches, he took the train to Calais July 21 with his wife and a group of friends to set up a vigil in the weather-bound Hôtel Terminus, beside the harbor. But he was filled with confidence in his little machine and in the outcome of the adventure.

Meanwhile, to replace the Antoinette IV, a new plane—Model VII —was rushed to Latham at Sangatte. (Models V and VI had already been consigned to other customers.) The latest model, employing a wing-warping device instead of ailerons, was considered so stable that it could almost fly without benefit of manual control. The cockpit was large, well reinforced, and padded; the firm's new and powerful eight-cylinder, 60-hp engine had been installed. All in all, the Antoinette VII was another sleek example of Levavasseur's fine craftsmanship—the realized dream of an engineer with the eye of an artist. It looked more than a match for the much slighter Blériot with its 25-hp, oil-spitting Anzani.

"Who Will Be Ready First, Latham Or Blériot?" queried the *Daily Mail* headline on July 23. Two French torpedo boats and the destroyer Escopette, as well as British destroyers, had been prepared for escort. The world's attention was again turned toward the French coast. Both the French and the British camps were nervously keyed up, each keeping an eye on the clouds as well as on the other. As usual everything depended on the wind—or the lack of it. "Waiting Aviators," read the July 25 headline. "Latham Ready To Start."

At 2:30 that morning Blériot was awakened by Alfred Leblanc, a balloon pilot who was a close friend of the family. Leblanc had important news for his colleague: the air was still and the stars were out. With some difficulty Blériot got himself out of bed, into his dark blue overalls, and onto his crutches. Anzani was already up; he had alerted the mechanics by firing four blanks from a revolver, and preparations were under way. The minutes sped by. At 4:00 the plane was brought out of its tent and wheeled into place; an elongated airbag rested in the fuselage, to keep it afloat in case of disaster. In the cockpit Blériot adjusted his skullcap with ear flaps, strapped on a life belt, and tied his crutches to the fuselage. It was still dark when the propeller started to turn. A small dog, barking angrily, ran into the whirling blades and was instantly killed. A bad omen, declared the pessimists.

A brief trial flight showed everything to be in order. The gasoline tank, holding 17 liters, was filled and the motor liberally doused with castor oil. The pilot was reminded to pump every three minutes to maintain the pressure. At 4:41, just as the sun broke the horizon. Blériot took off.

Latham had gone to bed late, leaving strict orders with Levavasseur to be called at 3:30 if the wind continued to abate. That he was literally caught napping was not his fault; Levavasseur, his sleep broken twice during the night, had dozed off until 3:50. A few moments later Latham heard the hum of his rival's engine—Blériot was already in the air.

It must have been an agonizing blow for Latham to realize, on looking out the window, that Blériot was on his way to England. His first plan was to set out immediately in pursuit. But in the short time necessary to ready the Antoinette, the wind rose again, putting flight once more out of the question.

Blériot, even if he had not a single instrument to guide him, carried luck on his side. The "waistcoat pocket monoplane," as the British termed it, quickly outdistanced the Escopette—steaming furiously in an effort to keep the birdman in sight—and disappeared into the hazy distance, out of sight and hearing. The aviator recorded his impressions as follows:

Below me is the sea, the surface disturbed by the breeze, which is now freshening. The motion of the waves beneath me is not pleasant. Within 10 minutes I have passed the torpedo boat, and I turn my head to see whether I am proceeding in the right direction, but I see nothing—neither the torpedo boat, nor France, nor England; I am alone. For 10 minutes I am lost, unguided, without compass . . . I let the airplane take its own course, and then, 30 minutes after I left France, I see Deal, which is far to the East of the spot I intended to land upon.

In a mounting wind, which affected his navigation, Blériot finally found the white chalk cliffs of Dover and, a few seconds later, the welcome landmark of Dover Castle. At 5:17 he came down to a rough landing in a little valley adjacent to the castle, breaking the undercarriage and one blade of the propeller. To those who rushed to greet the "Friendly Invader," much as if he were a man from Mars, the Frenchman presented a disheveled figure: eyes puffy and bloodshot, face wet with perspiration, hands glistening with oil. But in that short moment of time since he had risen from French soil, Blériot had become an international as well as a national hero. Not only had he beaten Latham, but he had vanquished the Channel—the timeless symbol of British impregnability. Like the "shot heard round the world" in 1775, it was a historic moment for mankind. "There are no islands any more!" was the cry.

The epic jump of 23½ miles in 36½ minutes, across an arm of the sea that had always been considered a formidable moat, shook the complacent British as few events could have done. "The day that Blériot flew the Channel," wrote Sir Alan Cobham, noted flyer of a later generation, "marked the end of our insular safety, and the beginning of the time when Britain must seek another form of defense besides its ships." The full implications of the exploit may not have been apparent in 1909—the Battle of Britain was still a long way off. But to the perceptive few it was full of portent.

Almost as incredible as the flight itself was the manner in which a French newspaper, Le Matin, adroitly reaped a lion's share of the credit for France. Although the Daily Mail had sponsored the event and arranged for its representative to be on hand at Dover to shower Blériot with congratulations, it was two French journalists, displaying a large French flag, who were actually the first to greet the aviator on English soil. Their coup sheds a revealing sidelight on the French knack for enterprise. As Michel Lhospice tells the story in his book Match pour la Manche, on July 20, 1909—the day after Latham's unsuccessful attempt and with Blériot's flight still in the offing—the editors of Le Matin met in Paris to consider the fact that the Channel race was fast making news—and newspaper readers. Maurice Bunau-Varilla, the publisher, a member of a family deeply interested in aviation, recognized that here was a heaven-sent chance to increase the paper's circulation. Consequently Charles Fontaine of the Paris staff, supported by Marcel Marmier, a photographer at the London bureau, were ordered to Dover to cover Blériot's anticipated arrival. Hurriedly reconnoitering the territory, they settled upon Northfall Meadow as the most likely landing site. On July 23 Fontaine entrusted to the captain of the Dover–Calais night boat a letter to Blériot, describing for his countryman the exact location of the field—which was graphically illustrated by two picture postcards and a map. Fontaine himself would be there to unfurl the Tricolor and wave it for a signal guide as soon as Blériot hove into sight.

By common consent among the journalists covering the competition at Dover, the word that one of the aviators had been sighted was to be spread by the foghorns on vessels moored at the wharf. Since no one knew what day would be favorable for a takeoff, some reporters took no chances and kept a nightly vigil for any advance news. During the night of July 24–25 anticipation was running high through the Lord Warden Hotel. One of the administrative vice-presidents of the Blériot establishment had arrived on the last boat from Calais and announced that he deemed it senseless to go to bed, as he expected to see the aviator over the English coast by 5 A.M. on the morrow. When, at 4:05, the wireless brought news of Blériot's imminent departure, the night porter roused all those still asleep—and within minutes the hotel was in a turmoil. While others ran to and fro, Fontaine and Marmier sped by motorcar to the spot they had picked out, bearing their country's emblem.

Their scheme worked to perfection. When Blériot climbed out of his cockpit, he was embraced with Gallic fervor by two Frenchmen, and photographed with the correspondent not from the *Daily Mail* but from *Le Matin*—together with the flag of France, prominently displayed. French self-esteem, never very far out of evidence, was at its height next day. *"Le Français Blériot Vient De Traverser La Manche En Aéroplane"* was spread in giant type across the front page of the Paris paper. The general reader might easily have received the impression that the whole affair had been arranged in France.

But the British had Blériot in tow moments after this exhibition of chauvinism. After breakfast at the Lord Warden Hotel, he was approached by three authorities from the customhouse, who—in the best tradition of their office—solemnly asked him if he had anything to declare. On answering in the negative, the flyer was granted clearance by an immigration officer in the following historic terms:

I certify that I have examined Louis Blériot, master of a vessel "Monoplane," lately arrived from Calais, and that it appears by the verbal answers of the said master to the questions put to him that there has not been on board during the voyage any infectious disease demanding detention of the vessel, and that she is free to proceed.

Blériot proceeded to London, where, through endless festivities, he was cheered, applauded, and praised. Never had the Entente Cordiale been so directly demonstrated. Under an improvised tent Model XI was inspected that first day by 12,000 persons irresistibly drawn to Dover Castle when the news of the crossing became known. For three days it was on view, free of charge, at Selfridge's Department Store in London, where it was seen by ten times that number. Gordon Selfridge was a businessman who believed it paid to advertise; he not only defrayed the expenses of transporting the plane to London but donated $1000 to charity to commemorate the occasion. Shortly afterward the spot where Blériot landed was consecrated by a bronze plaque—matched in 1911 by a monument at the starting point near Les Baraques. But the most gratifying material evidence of his success must have been the more-than-100 orders Blériot received for his monoplane within two days of the flight. The machine in which he made the crossing is preserved in the Musée des Arts et Métiers in Paris, while the broken propeller is a trophy in the bar of the Royal Aero Club in London.

While the furor was still fresh, Latham set out on a second try. Outwardly undismayed by Blériot's victory, he again pointed the propeller of his Antoinette toward Dover, on the evening of July 27. And this time he came within sight of his goal—only to be forced to alight once again on the water. The crowds awaiting Latham's arrival saw the machine glide down at a sharp angle, then throw up a column of spray as it hit the Channel. Once more the Antoinette motor had failed him. Possibly it was affected by a short burst of rain—though his family believe to this day that inexpert or careless mechanics were responsible for this double dose of trouble. Latham was thrown against the supporting mast between the wings and received a deep gash on the head. If he had not been rescued promptly, he might have lost consciousness, slipped into the water, and drowned.

Even this narrow escape, however, did not lessen his determination to fly the Channel. But the directors of the Antoinette company, having lost two valuable machines, were unwilling to risk a third. Latham would have to seek another outlet for his ambitions. He was to find it in the forthcoming meet at Rheims, for which all the available machines of the firm were being earmarked.

When Blériot returned, by boat and train, to Paris on July 28, he was met by the cheers of a hundred thousand delirious Frenchmen. The route that his automobile (with the irrepressible Fontaine up ahead flaunting his flag) followed to the offices of *Le Matin* was jammed as if a national holiday had been declared. Justly lionized on every side, the victor was engulfed in round after round of receptions. But of all the tributes Blériot received for his courage, and achievement, probably none pleased him so much as being made a chevalier of the Legion of Honor; the decoration was pinned on his breast by Paul Cambon, the French ambassador in London.

Blériot's flight had sent the Continent into a frenzy of excitement. It had showed that at last a European plane had been constructed that was worthy of the confidence of its producer; that flying over water was no more difficult than flying over land; and that the dawn of international air travel was at hand. It also served as an indication, perhaps not yet generally perceived, that the trend of accomplishment was about to turn from the United States toward Europe.

No sooner had the jubilation subsided than the next act of the drama commenced—this time near historic Rheims. With the aid of the local champagne industry and under the general direction of the Marquis de Polignac, a full week of flying events was organized in August in the interest of bringing together the world's greatest aviators. On the spacious plain of Bétheny, just north of the cathedral city, a mettlesome course was laid out—3750 meters for the main stretch, 1250 for the back. Pylons were erected to indicate the turns; tents and wooden hangars appeared to accommodate the various flying machines; a huge scoreboard filled the foreground; and a covered grandstand and enclosures were readied for an onrush of spectators. The spectators—practically none of whom had seen an aeroplane before—came by the tens of thousands; half a million was the lowest estimate for the week. Paris society, with the ladies in long dresses and picture hats, scarves, and parasols, streamed to the spot by train and touring car; 3000 Britons came from London by special excursion; 2000 Americans turned up to root for the single entry from the United States. Natives of the province of Champagne swarmed to the meet on foot and by bicycle, carriage, and cart. An excursion to Rheims for "La Grande Semaine d'Aviation"—with 200,000 francs in prizes to be taken—was a must, if not for novelty's sake then for a look at the notables present: President Armand Fallières, France's fan-bearded head of state; the Right Honorable Lloyd George, future prime minister of England; the genial British newspaper tycoon Lord Alfred Northcliffe, publisher of the *Times,* the *Mirror,* and the aviation-oriented *Daily Mail;* and high military officials of half a dozen interested countries. In the hotels and boarding houses every room was taken. Prices daily rose to new highs, in phase with the altitude records; concession-aires, sideshow barkers, peddlers, and photographers sprouted like dande-lions in the spring.

To crown the spectacle of man in flight a magnificent silver trophy was offered, ensuring the success of the world's first formal gathering of flyers. James Gordon Bennett, publisher of the Paris *Herald,* had already spon-sored an annual distance competition for spherical balloons, presenting for spoil the resplendent silver Coupe Internationale de l'Aéronautique. Now that mechanical flight was monopolizing the news, he had seized the moment to unveil a similar Coupe Internationale d'Aviation for establish-ment of speed records. Carefully designed by the fashionable Rue de la Paix silversmith André Aucoc, the trophy was in the form of a winged androgynous figure bearing aloft a likeness of the original Wright bi-plane. Each club or individual member of the Fédération Aéronautique Internationale could enter a team of not more than three pilots and their machines, supported by an equal number of substitutes. The trophy would go to the country of the winner. The victor nation would then be under obligation to play host to the competition the following year. Won three times in succession, the trophy would come into permanent possession of the prevailing country. For the first leg of the contest a purse of 25,000 francs awaited the successful pilot.

Bennett was an extraordinary figure in an extraordinary age. His dark hair parted in the middle, and sporting a heavy walrus moustache, he was noted for his wealth, his eccentricities, his social charm, his authoritarian will, and his vanity. The son of the multimillionaire founder of the New York *Herald,* he had exiled himself to France in 1877—ostracized by so-ciety for relieving himself while drunk in the fireplace of his fiancée's home. In Paris he later started the continental edition of the parent paper. In the words of Eric Hawkins, his managing editor for thirty-six years, Bennett "created the American-in-Paris image. . . . He was the most colorful American expatriate the French had ever laid eyes on. For forty years he enchanted Europe as publisher and playboy."

Bennett's enterprise was matched only by his originality: in 1870 he had sent Henry M. Stanley to Africa in search of the explorer David Livingstone; he had dispatched Marconi himself to New York to cover a yacht race by wireless; and he was the first to use the automobile to deliver newspapers to "distant" places, such as the bathing resorts along the Normandy coast. The newspaper magnate was soon known for the prizes and trophies he put up in yachting, motorcar, and balloon contests; his interest in racing on the sea, on land, or in the air was boundless. It was Bennett who provided a large part of the financial backing for the dirigibles and the Demoiselle of Santos-Dumont. Thus his sponsorship of the Rheims trophy stood as one in a long line of similar undertakings.

The early events on the program at Rheims were designed as entrées to whet the appetite for the *pièce de résistance:* the race for the Gordon Bennett prize, which was to be the culmination of the week's varied offerings. An initial entry list showed a total of thirty-five machines—twenty-two biplanes and thirteen monoplanes. The best-known pilots had flocked to the citadel of Champagne, lured by the promise of the prize money —and no less by the chance to exhibit their skill in the sheer sport of competitive flying. Here were Henry Farman, with his own new machine; Delagrange, with his Voisin model; Captain Ferdinand Ferber; and Count Charles de Lambert. Blériot himself, freshly enshrined in the hearts of his countrymen, was a heavy favorite; he was the champion of France, if not of the world—and his compatriots looked to him to gather still more honors for their nation. Latham also was a public idol for his pluck and perseverance. His gallant Channel tries had tremendously stirred imaginations, and his Antoinette was the most graceful plane in existence. Glenn Curtiss, the serious, lanky American, though still unknown abroad, commanded respect—although he puzzled most of those present by declining to show his colors in the first contests. With good reason, however, he held aloof: having already made one landing so

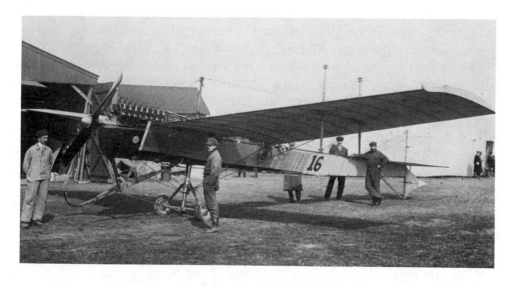

One of the Antoinettes at Rheims before the first Gordon Bennett cup race, 1909. (National Archives)

rough that he had sprained his ankle, he wanted to insure that his craft would be intact for the main international race.

Other flyers emerged from relative obscurity to become famous by the end of the week. Louis Paulhan, a cheerfully impecunious though clever mechanic who had been an early research associate of Ferber's, appeared with a Voisin won in a competition for construction of the best model of the big machine. He had borrowed the money for a motor and had taught himself to fly. At Rheims he gained the know-how for later triumphs. Henri Rougier, driver of fast racing cars and champion cyclist, also had learned just before the meet to pilot a Voisin, with which he was to glean his share of the prizes. George B. Cockburn, the only English entrant—and the first pupil of the recently opened Henry Farman school at Châlons-sur-Marne—had the nerve to enter the international meet with a machine newly bought from Farman, having yet made no solo flight longer than a quarter hour. Eugène Lefebvre, the chief pilot for the French Wright company, drew repeated salvos of applause for his antics in the air, which earned him the distinction of being named the first "aerial clown"—or stunt flyer. Etienne Bunau-Varilla swept past the grandstand waving his hat to the distinguished occupants, who waved back wildly. Esnault-Pelterie, handicapped by an injured hand, saw his original streamlined monoplane design meet with a measure of recognition at last.

Roger Sommer, the short, trim-bearded, slightly built son of a felt manufacturer from Mouzon in the Ardennes, made good use of the experience he gained at Rheims with a machine acquired from Henry Farman; for he was soon to construct a similar model under his own name that would set many new marks. This large but relatively light biplane was characterized by the extra length of its upper wing in relation to the lower. In 1911 he produced an elegant monoplane along the lines of Blériot's creations.

For a short time just prior to the meet, in fact, Sommer had held the duration record, as the result of a flight at Mourmelon-le-Grand on August 8 of 2 hours 21 minutes 15 seconds. Later, on October 28, 1911, he was to establish another kind of record by crowding seven passengers (all, admittedly, light in weight) around the seat, engine, and landing gear of his machine in a hop from Rheims to Mourmelon and back.

The roster at the Rheims tournament seemed comprehensive except for the absence of the Wright name. Despite the strong desire of Gordon Bennett to have them in the lists as champions of American aviation, neither of the brothers came to Rheims. Wilbur was busy in America; Orville was demonstrating for the Germans. It was a disappointment to the donor of the Coupe d'Aviation, not to speak of his many countrymen present, that American representation at Rheims was almost literally a one-man show—for Curtiss had only a couple of assistants, while the other contestants boasted large entourages of mechanics and self-appointed helpers.

Not unnaturally, French-made planes predominated. There were nine

Voisins and four machines each from the Farman, Blériot, Antoinette, and R.E.P. establishments; in addition a Bréguet biplane, in its first public appearance, was entered as well as a scattering of homemade models. Nonetheless half a dozen Wright or Wright-type machines (of which three actually flew), besides the one Curtiss, were reminders that the United States had been first with mechanical flight. It is worth noting that, with the exception of Bréguet, none of the names on the list of constructors survived among manufacturers after World War II. Like the illustrious Hispano-Suizas, the Isotta-Fraschinis, the Delauny-Bellevilles, and the De Dion Boutons of the early motorcar age, such favorites as the Antoinette, and the Henry Farman, the Voisin, the Blériot, and the Wright are but memories of another day. Unable to attract orders from the military and with only sporadic sales to civilians, these famous manufacturers could not produce at a profit and were forced ultimately to close their doors.

When the Rheims meet opened on August 22, 1909, rain was descending in sheets. The flying field was flooded, cars were bogged down, and planks had to be laid at strategic points for pedestrians to cross. No worse weather in which to try one's wings could be imagined. From a mast at the judges' stand flew a black flag, signifying that flight was impossible. As the day progressed, however, a small group braved the gusts and the downpour to try for a place on the French team in the Gordon Bennett race. Lots were drawn for the order of start, with priority going to one of the red monoplanes from the R.E.P. camp; but no amount of effort could get this machine into the air. Paul Tissandier of the Wright school (whose father, Gaston, was one of the great balloonists of his time) was the first to be airborne—a trial of merely a minute. Blériot, in his Channel-crossing Model XI, was up next—but descended quickly with a clogged carburetor. Latham, sporting the number "13" on wings and fuselage, gave the superstitious cause to feel justified when his Antoinette was forced down after 500 yards. Lefebvre was unable to complete two laps, while Captain Ferber—flying under the name F. de Rue (after his family's estate at Rue, near Lausanne, Switzerland)—strove ineffectually to raise his heavy Voisin out of the muck. A more disappointing run of performances, especially as speed was the criterion, could hardly have been given. Those who did extricate themselves from the soggy field had mud clinging to wheels and tail skids, streaks of dirt on the undersides of wings.

Since Lefebvre and Blériot had made the best showing, such as it was, they were chosen to represent France in the big race. That afternoon, when elimination heats were resumed under moderating skies, third place went to Latham; Tissandier, de Lambert, Paulhan, and Sommer were designated as reserves, in that order.

In the speed contests that followed, the public had its first chance to argue the relative merits of biplane versus monoplane. To the general surprise, in view of the monoplane's supposed advantage in this respect,

Louis Blériot in his model XII monoplane at Rheims. (Paul Nortz Collection)

Charles de Lambert in his Wright biplane at Rheims. (Paul Nortz Collection)

the Wright pilots made a clean sweep. First, second, and third place went to Tissandier, Lefebvre, and de Lambert, respectively—each flying a staid and unspectacular Wright biplane, with its front elevator—a trip which was, as someone said, "like going for a ride on your front porch." Fastest lap on that difficult inaugural day was credited to the dexterous Lefebvre, who captured a time of 8 minutes 58⅕ seconds for the 10-km course.

As the meet continued, the number of operable machines rapidly dwindled. As in the tourneys of the Middle Ages, where the lance of one after another visored knight was splintered and his armor was broken, the preliminary spins and trials took their toll of wood and muslin fabric. Even a minor mishap often put a plane out of the running. Balky engines declined to be coaxed to life and required hours of adjustment. Some pilots, indeed, never left the ground at all. On one occasion Glenn Curtiss, rounding the course in a tryout, saw "as many as twelve machines strewn about the field, some wrecked and some disabled and being slowly hauled back to the hangars by hand or by horses."

Nevertheless records were regularly set and then broken. In the altitude contest Latham rose to the breathtaking height of 155 meters—the limit that the judges would allow on the basis of his barograph recording, although some were sure he had gone higher—while more conservative flyers did not venture to even half this elevation. In the long-distance category as well, the virtuoso of the Antoinette gave an exhibition that would never be forgotten. Riding his temperamental mount with the tenacity and *sang froid* of a broncobuster, he refused to be deterred by the sight of threatening thunderheads and flew 154.5 km through a series of nasty rain squalls—a mark that stood till near the end of the meet. So heroic was the combination of man and machine that some of Latham's more emotional female admirers wept openly when they saw the Antoinette towed back to its hangar with a bent propeller. Levavasseur, possessed of a stronger constitution, watched imperturbably; he never took his eyes off his beloved Antoinettes in flight and never lost his sense of humor if things went wrong.

New marks for speed, over one or more laps, were recurrently hung up. Henry Farman discarded his troublesome Vivinus engine of Belgian make, installed one of the new French rotary Gnômes, loaded two passengers into his *cage à poules,* and carried them to a new record, winning the Prix des Passagers. At one point no fewer than seven machines were in the air at the same time. And then, on August 27, Farman set the world agog by reeling off the incredible distance of 180 km in a flight of 3 hours 4 minutes 56⅖ seconds—a flight that finished in semidarkness. That left Latham in second place as far as distance was concerned, and Paulhan third with 131 km. It not only gave Farman the Grand Prix but gained him the largest share of the prize money—computed at 60,000 francs. Latham received 48,650 francs, Curtiss 38,000 francs, Blériot 12,000 francs, and Paulhan 10,000 francs. Miraculously no serious injuries and no fatalities marred the eventful week.

The closing day brought weather the exact opposite of the opening: sunny and hot. A crowd of a hundred and fifty thousand turned out for the climax. The Gordon Bennett race, though three nations were represented, looked like a duel between the United States and France—between the Curtiss biplane and the Blériot monoplane, both of which were thought to be fast. The sole British starter was not given much chance. Two other places had been reserved on the British team, but they were not filled; nor did the expected entry of three Italians and one Austrian materialize.

For all his fads and foibles, James Gordon Bennett was modesty itself in print; for while everyone spoke of the "Gordon Bennett cup," it was never so called in the Paris *Herald.* "News, news, news. Names, names names—that's what makes a paper," was his credo—but that did not in-

clude the use of his own name. This diffidence did not detract in the least
from the personal fame Bennett enjoyed as a patron of flight—nor from
the sales of his aviation-minded newspaper—and his name was on everyone's lips when the speed classic was called on August 28. Attendance that
day swelled to climatic proportions; it included Ambassador Henry
White of the United States and Mrs. Theodore Roosevelt, whose husband
had recently finished his second term as President. Anticipation mounted
with the summer heat; the grand finale was certain to be a thrilling
climax.

Getting America's entry to the starting line had been a touch-and-go
affair. Curtiss had won the second leg of the *Scientific American* trophy on
July 17 by flying 24.95 miles in 52.5 minutes at a speed of 28.51 m/hr
(not counting turns) on the 1.3-mile triangular course at Mineola, Long
Island. As soon as he had been notified by the Aero Club of America, on
the strength of this performance, that he had been selected to represent
the United States at Rheims, he had begun the construction of an eight-
cylinder V-engine of 50 hp—double the power he had hitherto used—
which constituted his main hope of winning. Working secretly day and
night in the Hammondsport shop, he finished the engine barely in time to
catch the ship that was to take him to France. There was no chance for exhaustive tests; the new motor, hurriedly mounted on a block, was given a
short run, packed, and rushed to the dock, along with the dismantled plane.
Faced with bureaucratic delays at the French customs on arrival, Curtiss
persuaded the officials—perhaps themselves infected with the current flying fever—to let him check his three crates as personal luggage. He
reached Rheims only hours before the required qualifying rounds.

On the day of the race, however, Curtiss was ready. His Golden Flyer
(successor to the June Bug) was a sturdy pusher biplane with wings of
yellow oiled silk, front elevator and tail, ailerons between the wings, tricycle landing gear, and engine and radiator mounted above and to the rear
of the pilot. Picked over the Wright biplane, with its twin propellers and
greater wing span, not to mention an established reputation, the Curtiss
design seemed to promise greater speed. Curtiss's first flights over foreign
soil appeared to confirm the wisdom of that choice. A number of refine-
ments had been carried out at the last moment. To reduce weight and
wind resistance, the large gasoline tank had been replaced with a smaller
one, the capacity of which appeared barely adequate for the course. Over
and over again Curtiss debated the selection of a propeller. He tested the
wiring by lifting the machine at the corners, and methodically checked the
engine with his mechanic.

When the race was officially declared open at ten o'clock on that blaz-
ing Saturday morning, the American machine was tuned up and waiting
well ahead of the others. At ten thirty Curtiss was in the air on his one
permitted test flight. After two bumpy circuits, which he described as
"difficult but not dangerous," he sent word to the judges that he was

ready to make his official try for the cup. What followed is best told in his own words:

I climbed as high as I thought I might without protest, before crossing the starting line—probably five hundred feet—so that I might take advantage of a gradual descent throughout the race, and thus gain additional speed. . . . I cut corners as close as I dared and banked the machine high on the turns. I remember I caused great commotion among a big flock of birds which did not seem to be able to get out of the wash of my propeller. In front of the tribune the machine flew steadily, but when I got around on the back stretch . . . the air seemed fairly to boil. The machine pitched considerably, and when I passed over the "graveyard," where so many machines had gone down and were smashed during the previous days of the meet, the air seemed literally to drop from under me.

But Curtiss kept the throttle wide open and completed the two laps in 15 minutes 50⅗ seconds at an average speed of about 47 m/hr. That was the mark his competitors would have to beat.

Next aloft was Britain's one hope. George Cockburn, however, flew in a hopeless cause—for England in 1909 was behind in aviation. "Colonel" Samuel Franklin Cody, a flamboyant figure of Texas origin though no relation of Buffalo Bill, had made the first officially recorded flight in Great Britain on October 16, 1908, at the Farnborough Balloon Factory. His machine—designated Army Aeroplane No. 1—covered 1390 feet in 27 seconds before it crashed. In August 1912, Cody, who had become a British subject by naturalization, was to achieve further fame by winning the British Military Aeroplane Competition with a towering biplane of his own construction—sometimes referred to as "The Flying Cathedral." A few other inventors, including Geoffrey de Havilland, Frederick Handley Page, and A. V. Roe, were beginning their experiments. But aspiring pilots—and there were not many—looked to the Wrights, Voisin, or Henry Farman for their machines, and to France for training and experience. For instance, Moore-Brabazon, holder of British license No. 1, learned to make short hops at Issy-les-Moulineaux and Châlons-sur-Marne in a plane built for him by Gabriel Voisin. His second Voisin (the one intended for Henry Farman)—the Bird of Passage—he took to Shellbeach (Leysdown) on the Isle of Sheppey; there, between April 30 and May 2, 1909, he succeeded in making the first flight by a British-born Briton on his native heath.

As the first representative of his country to take part in an international meet, Cockburn did his best; but his much slower Farman was in any case no match for the speedy Curtiss. The Englishman was unable to complete even one full lap of the course. On descending he ran into a haystack—an inglorious end to Britain's chances.

The French thus had the field to themselves in their effort to beat Curtiss. When Latham took off, to rousing cheers from the stands, nothing was considered impossible. A rumor (grossly unfounded) had spread that the Antoinette was capable of 60 m/hr and would offer the strongest competition to all comers. But not even Latham's expert manipulation could drive this mechanical beauty to improve on the time of the American: it finished in 17 minutes 32 seconds—nearly 6 m/hr slower—and was automatically out of the contest.

Even more discouraging to Frenchmen was the showing of the popular Lefebvre in his Wright. Lefebvre had performed consistently well in the other events; but the Gordon Bennett contest called for something more than consistency. A few seconds clipped off at the turns, an extra thrust from engine and propeller, a more daring bank than usual before going into the straightaway, were needed in a race where every mile and every moment counted. The Wright was not equal to the task; its pilot could not extract more than 35 m/hr from the craft, and, with a score of 20 minutes 47⅗ seconds, it never was a serious contender. If the day was to be saved for France, it was up to Blériot.

Toward four o'clock, pulses quickened as the mechanics trundled out Model XII. This was a big monoplane for its time, with a brand-new eight-cylinder, 60-hp, water-cooled E.N.V. engine weighing 287 pounds —a British type then known for its reliability in speedboats—mounted directly beneath the wings. This advanced two-seater, aileron-equipped, high-wing job (which its inventor had abandoned in favor of the Model XI for the Channel crossing) carried a four-bladed, chain-driven propeller at the leading edge, geared to turn at a slightly slower speed than that of the engine. Blériot had spent the whole day plotting and considering one combination after another of plane, propeller, and engine. Now, with the American threat a real one, he felt he had the answer; and to the on-lookers, as the most powerful machine at Rheims roared into action, it seemed that he could not fail.

From the moment he signaled the mechanics to let go, and the machine rocketed into the air after the briefest of runs, Blériot's most recent creation—and, with the new engine, his fastest—gave the impression of utter invincibility. To the end of the first lap he was clocked at 10 seconds faster than Curtiss, and came close to an unprecedented mile a minute on the backstretch. But impressions could be deceiving. As the French flyer finished the second lap and sped in to a perfect landing, his friends prepared to congratulate him—and even Curtiss himself was ready to concede his defeat. When the Stars and Stripes was hoisted at the judges' stand and the band struck up the American national anthem, the crowd was dumbfounded. Blériot's elapsed time was found to be 15 minutes 56⅕ seconds—6 seconds longer than that of Curtiss. America had won by a slim margin.

From every American throat burst a prolonged cheer, drowning out the stunned response from French patriots. It was a day of glory not only for the United States but for the allegedly slower biplane. As the Paris *Herald* observed next morning, the race had "rehabilitated" the biplane's image. While the birdlike monoplane had a mass appeal because it seemed ideally to represent artificial flight, "the American aviator proved that the biplane not only possessed qualities of carrying weight and undoubtedly of superior stability, but that, if need be, it can develop speed equal to, if not superior to, its smaller rival."

The day's thrills, however, were not over. For an added attraction, two of the region's leading champagne firms had sponsored a three-lap race. Blériot again took to the air, determined to prove that his machine was after all capable of a winning performance. But suddenly he was compelled to land. To the horror of the crowd, the plane burst into flames —an overheated fuel pipe had spilled gasoline on the exhaust. Blériot was able to walk away from the wreck with nothing worse than injuries to his arms and face, while Model XII burned to a skeleton—another in the series of accidents that dogged his career.

Aero Club officials in America were in a state of justifiable euphoria when, after exhibitions in Germany and Italy, Curtiss finally came home. Receptions without end awaited the winner: luncheons, dinners, "Welcome" arches in electric lights at Hammondsport—with banners, speeches, fireworks, and a carriage drawn by fifty men to meet him at the station on the arrival of a special train. With the fastest aeroplane in the world, it was confidently predicted that America would win again the next year—that the trophy would be staunchly defended and ultimately remain on this side of the Atlantic.

The Wright brothers, too, added to the general enthusiasm. At the Hudson-Fulton Celebration in New York during the last of September and the first week in October, Wilbur enthralled millions of spectators —capping his performance with a venturesome trip up the gusty Hudson from Governor's Island to beyond Grant's Tomb and back, a total of nearly 21 miles. A million-dollar Wright company was formed shortly afterward by a syndicate that included the names of such famous and confidence-inspiring capitalists as Vanderbilt, Gould, and Belmont.

Unfortunately the optimism did not last. The next Gordon Bennett race was to go to a foreigner, in a foreign machine. The Curtiss triumph represented the first victory for a biplane; it was also the first—and last—for an American-made machine. The French over the ensuing years were to be amply compensated for their defeat at the initial tourney. The victory of Curtiss marked the end of America's ascendency in the years preceding World War I: it was the only instance during this period in which a plane produced in the United States competed successfully in an international tournament.

Financially, technologically, and socially, the display at Rheims was a sensational success. Some ten different makes of machine were on view, six of which—the Wright, Voisin, Henry Farman, Curtiss, Blériot, and Antoinette models—were offered for sale. All except the Voisin had well-developed control systems, and all were capable of speeds of greater than 40 m/hr. More than 120 takeoffs were made, of which 87 produced flights of at least 3 miles—a practical demonstration of the performance capabil-

Four French pioneers at the 1909 aviation meet at Rheims. (Paul Nortz Collection)

Louis Paulhan

Eugène Lefebvre

ities of the different models. The price at that time varied with the make. The Wright—the most expensive—sold for 30,000 francs; the Antoinette for 25,000; the Farman for 23,000; the Voisin for 12,000; and the economical Blériot XI for 10,000.

In addition to being a gay period piece—an example of the Belle Epoque at its colorful best—the week was a watershed in the history of aviation: an aviator who had flown "before Rheims" was distinctly a veteran compared with one who learned later. Rheims showed the world what strides had been made and how capable men had become in handling their machines. It showed, too, how much they had still to learn—and how easily their craft could be wrecked or damaged. The tournament set the pattern for the many aerial meets to follow in both Europe and the United States. It provided an incalculable stimulus for the design and production of aeroplanes. And lastly, the Rheims competition helped turn popular attention from the ground to the air. When the lists closed, on November 30, 1909, for the Grand Prix of the Automobile Club de France, only twelve of the necessary forty-five entry applications had been received—and the contest was canceled as being no longer of sufficient public interest.

Until the end of December 1909, the Aéro Club de France issued licenses without examination to pilots who had actually flown; sixteen such licenses had been granted up till then. On January 1, 1910, the list was

Roger Sommer

Robert Esnault-Pelterie

published for the first time, arranged in alphabetical order to avoid the slightest suggestion of favoritism or judgment of relative ability. The sixteen were as follows:

No.	Name
1	Louis Blériot
2	Glenn H. Curtiss
3	Léon Delagrange
4	Robert Esnault-Pelterie
5	Henry Farman
5a	Ferdinand Ferber [number assigned posthumously]
6	Maurice Farman
7	Jean Gobron
8	Charles de Lambert
9	Hubert Latham
10	Louis Paulhan
11	Paul Tissandier
12	Henri Rougier
13	Alberto Santos-Dumont
14	Orville Wright
15	Wilbur Wright
16	Etienne Buneau-Varilla

From that point on, the club's policy changed. The alphabetical order was given up, and regular tests were required. Thus, Alfred Leblanc came after Bunau-Varilla, with license No. 17; Julien Mamet, a Blériot pupil, received No. 18; and so on. The first sixteen, however, were the names on everyone's tongue—the undisputed leaders in an activity that had been virtually nonexistent twelve months earlier.

As Lloyd George remarked in an interview, "Flying machines are no longer toys and dreams; they are an established fact. The possibilities of this new system of locomotion are infinite." A new age was in fact dawning—heralded by the first Salon Internationale d'Aviation in November 1909 in Paris. It was already obvious that France had forged into the lead—in number of licensed pilots, in new records conquered, in establishment of aeroplane and motor factories, and in competitions that resulted in increased skill as man pushed the boundaries of his experience further. The American contribution would continue; but the pace would be set by Europe.

CHAPTER FIVE

The Tournaments Begin

If America was the cradle of aviation, France was the progressive nursery school. Nothing could have provided more incentive to fly than the spirit of achievement born at Rheims. If, as Levavasseur asserted, man should be able to soar better than the birds, clearly the time had come for him to try. In Europe—much more readily than in the United States—the challenge of the new sport was eagerly accepted.

Like the buzzing of giant insects, a strange, new sound was heard in the land. Experimenters and tyro constructors, turning open spaces into flying fields, filled the air with the raucous drone of their motors in the unthrottled exercise of power. A rash of aerodromes appeared in the vicinities of the big cities. Around Paris alone, in addition to Issy-les-Moulineaux, one could see flights at Vincennes, Saint-Cyr, Villacoublay, Juvisy, Buc, and Etampes. Racetracks, parade grounds, sporting clubs, cow pastures—in fact any level piece of ground not hemmed in by trees or telegraph wires—became potential flying fields. For a town of any pretensions at all, it was a matter of civic pride to promote exhibitions of flying. Country folk tilling their fields might find the peace of a perfect day unexpectedly shattered by the roar of an engine; gazing upward open-mouthed, they would behold the spectacle of man and machine overhead. Like blossoms that burst in the sudden warmth of spring, aerial events were occurring everywhere.

The assembly at Rheims was the forerunner of many other gatherings—in Italy, France, England, Germany, and the United States. Promoters, scenting the possibility of profit, rushed in while the propellers were still turning to sign up every aviator within reach. During the period of September 8 to 20, 1909, a multinational bevy of flyers came to Italy, under the auspices of the Società Italiana di Aviazione, to compete in a Circuit of Brescia, on the sunny northern plain of Lombardy. Including Blériot and Curtiss—the latter's laurels still fresh after the Gordon Bennett race—they easily outflew the budding Italian pilots. Only Wilbur Wright's pupil, Lieutenant Mario Calderara, was able to hold his own; with a Wright machine acquired by the Club Aviatori di Roma, Calderara completely outclassed such local oddities as Franz Miller's *aerocurvo* (the product of Turin's first aeroplane and motor constructor) and the Pasotti biplane, an invention of the engineer Rovida. Curtiss was at his best at Brescia; he won first prize, and it was to his championship biplane that the poet d'Annunzio entrusted himself for his ecstatic first ride through the air.

It was not altogether smooth going, however. Landings were hazardous on the rough, uneven ground; and once, when the announced flight program was canceled, the crowd became so angered and unruly that troops had to be called out to quell the disturbance. But the meet started Italy on the road to recognition as a power in aviation.

Also on the heels of Rheims, an "October Fortnight" was organized at Juvisy's Port-Aviation, some twelve miles south of Paris. In this demonstration, and purely for sport, Count de Lambert took off on October 18 and circled the Eiffel Tower in a flight that lasted 49 minutes 39⅗ seconds. It was the first flight over the city—an unforeseen climax to a series of exhibitions that had stirred Parisians to the core. At Juvisy such enormous crowds engulfed the suburban trains that the system finally broke down, stranding thousands of passengers and resulting in a stampede that left scores seriously injured.

In England during October, flight meets were held simultaneously at the seaside resort of Blackpool and the midlands manufacturing town of Doncaster, in what proved to be an unedifying display of commercial rivalry. Though the iron of public interest was still hot, there was scarcely room for two shows; and the autumn climate in the British Isles was not so conducive to flying as was that of Italy. Under the sponsorship of the British Aero Club, the Blackpool exhibition starred Henry Farman, Louis Paulhan, and Latham—the first two taking turns on the same machine. Farman's flying seemed less than brilliant, for he hugged the ground at a height of only 8 feet in monotonous circuits of the course. Other Frenchmen—Henri Rougier, with his Voisin, and Maurice Tétard, with a Sommer-built biplane—were severely handicapped by the weather. Paulhan, the tenth man in France to receive a license, showed more versatility than most; but Latham's performance was unforgettable. Hailed as "The Storm King" at Rheims, he again gave the lie to those who said aeroplanes were useless in a wind. To the alarm and suspense of his friends ("Come down, you splendid fool!" screamed one), he bucked a "gale" estimated at 30 m/hr, with gusts of 40 m/hr. Again the Antoinette proved itself a natural wind fighter, flying literally to a standstill at one point, and completing only 4 miles in 11 minutes—one of the slowest flights yet made.

At Doncaster the principal performers were Léon Delagrange, Hubert Leblon, and Roger Sommer—a trio of hardy and determined pilots. But the weather there was even worse, and less experienced aviators hardly emerged from their sheds. Those present but showing prudence included such early amateurs as Emile Bruneau de Laborie, with a Henry Farman; Paul de Lesseps and G. Prévoteau, with Blériots; Emile Ladougne, with a Goupy biplane; and Louis Schreck, with a Wright. England was represented by Captain E. M. Maitland, who operated a Voisin with which he was just beginning to fly. Also on the scene was Hélène Dutrieu, a thirty-two-year-old French sportswoman known as "The Girl Hawk," who had learned the hard way—cracking up in a Demoiselle—but who was soon to break world records for women with a Henry Farman machine. The daily fare of squally weather forced pilots into disappointing performances, and the public for the most part remained unimpressed.

Things were better in Germany for those who had scattered for further fame and fortune after Rheims—and for some newcomers too. The first German meet was held at Berlin in September. In a sequel organized at Frankfurt am Main, the wealthy Belgian baron Pierre de Caters, in a specially built Voisin with a 60-hp Gobron engine, won a $10,000 first prize. Smaller meets were held at Cologne and at the historic watering town of Spa, across the Belgian border.

As an afterthought, to make up for the embarrassing lack of applause in England, arrangements were made for Louis Paulhan to show Londoners what flying was like. In November, at the Brooklands motor racetrack near Weybridge, he fulfilled a flying engagement with such cool skill and daring that his reputation was made. The demonstration attracted attention to Brooklands as a possible future aerodrome. It was presently converted to a flying field and developed into a famous center of British experimentation.

From January 10 to 20, 1910, Paulhan took a leading part in America's first international air tournament, held at Los Angeles under the management of Dick Ferris, the actor. At the old Dominguez ranch, a few miles south of the city near Compton, twenty-five thousand people packed the grandstand; thousands more milled around the field, watching imported Blériots and Farmans compete with domestic products. Balloons and dirigibles floated lazily above. Pitted against the Frenchman and two nascent fellow flyers, Didier Masson and M. Miscarol, was a team from the Curtiss school, headed by Glenn Curtiss himself: Clifford B. Harmon, a millionaire balloonist and sportsman; Charles K. Hamilton, a former trick bicycle rider, parachute jumper, and glider and dirigible operator; Charles F. Willard, the first man taught to fly by Curtiss; and Frank Johnson.

To test the qualities of a plane and the proficiency of its pilot, the aerial meets of that time programed different events demonstrating various characteristics: endurance, altitude, quick starting, accurate landing, bomb dropping, weight carrying, quick turning, cross-country flight, fast speed, slow speed. Quick starting was defined as "the distance covered before leaving the ground with motor running" or as "time from first explosion to time of leaving the ground." The quick turn was one complete circle in the air.

Many of these events helped to accelerate the progress of aeroplane design. Competitions for fast speed led to new methods of construction that

Souvenir program for America's first international aviation meet.
(National Air and Space Museum)

lessened air resistance; endurance contests spurred development of more dependable motors; altitude marks were set one after another as lessons were learned in aerodynamic design. To make a quick turn was impossible with the inherently stable Voisin, while the Wright machine, with its flexible wing tips, was highly maneuverable; the difference thus pointed up soon brought the "stable" plane into disfavor. Nothing could beat the biplane for all-around performance at Los Angeles. Curtiss broke the world quick-start record by taking off in 6⅖ seconds after a run of 98 feet; he also set a new speed mark for pilot with passenger, at 55 m/hr. Hamilton made the slowest lap, at 26.83 m/hr. But on January 12 Paulhan took top honors by soaring to the unprecedented height of 4164 feet—promptly certified by the Aero Club of America as several hundred feet above the previous world record—a feat that earned for the Dominguez field a lasting fame in the chronicle of aviation.

Fortune continued to favor Paulhan. On January 20, the last day of the meet, he won the endurance prize with a 64-mile flight in 1 hour 49 minutes 40 seconds. He also made a number of adventurous trips over the countryside, during one of which he was cheered on by a crowd that tried to follow him along the rural roads on horseback, by motorcycle, and by automobile. On his return, spectators stormed out of the grandstand, hoisted the flyer on their shoulders, and bore him off in triumph. His winnings came to $10,000.

Paulhan contrived in other ways to stay in the limelight. On January 19 he carried as a passenger Lieutenant Paul Beck of the U.S. Army, who practiced dropping weights with the aid of a crude bombsight—the first hint of aviation's potential in warfare. The French aviator went on to other adventures in the United States, including a smashup on February 4 at the Overland Park racecourse in Denver—in which he was providentially hurled into a bank of soft snow—and a bout with the Circuit Court of New York, which had granted an injunction restraining him from the use of his machine in exhibition flights in the United States. (The ailerons of the Farman biplane, it was charged, infringed on the Wright brothers' patents.) But Paulhan never forgot Los Angeles, and Los Angeles never forgot Paulhan. In 1960, at the age of seventy-seven, he was among the invited guests on Air France's inaugural nonstop jet from Paris to Los Angeles—a fitting commentary on fifty years of flight progress.

Migration to the calm of warmer climes than that of England and northern France was attractive, as well as lucrative, to Europe's birdmen in winter. In February 1910, a week of racing, twentieth-century style, was organized at Heliopolis, Egypt—"City of the Sun." Visitors to that convocation saw the winged motifs of deities on ancient walls come to life, as it were; the mythological days when "only the gods could fly" were forever past.

Competitors and spectators alike were well rewarded for their sojourns. Henri Rougier garnered $18,200 in prizes with his Voisin, while René Metrot—another Voisin pilot, who had received his license but a

month before—netted winnings of $12,000. Leblon, with his Blériot, an-
nexed $3,200; Jacques Balsan, a former sportsman-balloonist and member
of a wealthy family of textile industrialists, took in $1,700 with a Blériot;
Frederick van Riemsdyk, a Dutchman flying a Curtiss-type biplane, made
$500. In her first public appearance, Baroness Raymonde de Laroche, an
intimate friend and pupil of the Voisins, represented her sex with panache.
The raven-haired Raymonde, who affected a modish toque hat with
aigrette plume when not in flying togs, became—on March 8, 1910—the
first woman in the world to win a pilot's license.

As spring came to the Mediterranean, the scene shifted to Cannes and
Nice. Aeroplanes were a welcome diversion for resort society, and pilots
were lionized the length of the fashionable Riviera. Opening the events on
March 27, Léon Molon—his license only two months old—crashed on
the seventh round of the aerodrome at Cannes; his Anzani-powered Blériot was a hopeless wreck, but Molon escaped with only minor injuries.
On April 3 the Russian pilot Prince Nicolas Popoff winged his way in a
Wright to the Lérins Islands and back in 18 minutes 20⅗ seconds to win
the Prix du Voyage—a feat which caused him to be mauled by enthusiastic crowds on his return and congratulated in person by Her Imperial
Highness the Grand Duchess Anastasia of Mecklenburg. The excitement
carried over into "Aviation Week" at Nice—highlighted by a race across
the water to Antibes and back on April 24. Here Latham's propensity for
falling into the sea was demonstrated anew: while rounding the light-
house, his motor stopped, and again he had to be rescued from his floating
machine.

The purse at Nice was also a rich one. Michel Efimoff, a Russian who
had taken his license in a Voisin at Châlons in February, was the high
scorer; he reaped prizes equivalent to $15,510—a victory which led to his
appointment as instructor at the Russian School of Military Aviation, at a
reported annual salary of $15,000. Latham was next with $12,110, followed by Charles Van den Born of Belgium, who earned $5,440.

Elsewhere also, prize money continued to flow freely, tempting aviators
to disregard the risks involved in its pursuit. Fame awaited in every category. To take only one example, the scientific journal *La Nature* put up
10,000 francs for the first cross-country flight of 100 km in 2 hours or
less. This challenge was met on April 3 by Emile Dubonnet, son of the
celebrated winemaker, who piloted a light, easily controlled monoplane
designed by the Tellier brothers and powered by a 35-hp Panhard-
Levassor motor. Despite the contribution made by the Tellier craft to
monoplane development, it soon thereafter disappeared in a welter of
competing designs.

Nobody had believed, two years earlier in 1908, that the biggest prize
of all—the *Daily Mail*'s dazzling offer of £10,000 for a flight from London to Manchester within 24 hours—would ever be won. It was in fact
openly mocked by the rival *Star:* "Our own offer of £10,000,000 to the
flying machine of any description whatsoever that flies five miles from

London and back to the point of departure still holds good. One offer is as safe as the other." The magazine *Punch* joined in the laughter with an offer of £10,000 to the first "aeronaut" to fly to Mars and back within a week.

A good-looking, magnetic English sportsman, Claude Grahame-White, decided to try for this lavish prize. But first he had to learn to fly. One of Blériot's first pupils, Grahame-White—a dealer in motorcars—had stood on the sidelines at Rheims, captivated by the spectacle of flight. After lessons at Issy in a duplicate of Blériot's ill-fated Model XII, he took his two-seater to the increasingly popular flying center of Pau, where it was wrecked—the last of its kind—in an attempted landing with Blériot himself at the controls. Grahame-White's next step was to open his own flying school. Shortly afterward he learned at Châlons how to handle a Henry Farman biplane—and with this machine he entered the contest for the *Daily Mail* prize.

A first attempt, on April 21, was cut short by engine trouble after the English flyer had covered over 100 miles and his *cage à poules* had been blown upside down by a squall where he had alighted. On the second start, with repairs effected, he was challenged by Louis Paulhan—in one of the most dramatic episodes on record.

At sunup on April 28, the wind seemed too boisterous for flying that day. Like Latham waiting at the Channel, Grahame-White went back to sleep in the hotel near his point of departure at Wormwood Scrubbs. At 6:00 P.M., he was awakened with the news that Paulhan, piloting an identical Farman, had taken off from Hendon at 5:31—bound for Manchester. Although darkness was falling, Grahame-White hurriedly got his own machine into the air and reached the town of Roade (a distance of 60 miles) before he was compelled to come down for directions. To cut Paulhan's lead, the resourceful Englishman resolved to fly by night—something no one had done before. At 3:00 on the morning of April 29 he was airborne again, guided by the headlights of an automobile and the outline of a freight train moving toward the same destination. But his daring ploy was to no avail: a recalcitrant engine, abetted by a heavy wind, brought him down at Polesworth, 107 miles out of London.

Paulhan meanwhile, escorted by a special train of the London and North Western Railway, had made a landing in deep twilight near Lichfield. At 4:00 A.M., in the faint light of morning, he left on the last leg, fighting a wind which he later described as "dreadful." Contrary to the ordinary condition of early-morning calm, he was "hit by sudden, vicious gusts"—which, because of its long forward elevator, caused the Farman to pitch mercilessly. Describing the flight, Paulhan related: "My machine rose abruptly and then dropped again so rapidly that I was almost jerked out of the driving seat. My arm ached with operating the control lever. I soared up to more than a thousand feet in the hope of finding quieter air, but still the wind pursued me." Nevertheless the aviator reached Manchester without mishap. He was well within the time limit of 24 hours; he had made only one of the two permissible stops; and he had set a record of 183 miles in 242 minutes of flying time.

At a gala luncheon in the Hotel Savoy in London, Paulhan received the £10,000, in gold—ten times the sum won by Blériot for crossing the Channel. It was a stirring occasion: an exhibition of sportsmanship well calculated to strengthen the good diplomatic relations between France and England that prevailed at the period. Ambassador Jules Cambon of France was invited to make the presentation, and the prize reposed in a casket of gold on which the British and French flags were emblazoned in emeralds. For good measure Paulhan took another £5,000 from the generous *Daily Mail* for the greatest number of flights during the twelvemonth period ended August 14, 1910.

Across the ocean, on the banks of the Hudson, another substantial sum of money awaited—$10,000 for the first flight between New York and Albany. The New York *World* had announced the prize on January 31, 1909; the passage was to be a feature of the Hudson-Fulton Celebration, duplicating by air the momentous voyages of Henry Hudson's Half Moon in 1609 and Robert Fulton's Clermont in 1807. Two attempts by dirigible —one by Thomas S. Baldwin and one by George T. Tomlinson—had ended in failure. Moored in great tents on New York's river front behind a high fence painted with the sign "Hudson-Fulton Flights," the dirigibles were scarcely capable of more than local excursions. On October 19, realizing the difficulties of the project at that stage of aircraft development, the *World* extended its offer, to be valid in either direction until October 10, 1910.

Glenn Curtiss accomplished the trip in his biplane on May 29—the first long-distance aeroplane journey in the Western Hemisphere. With a flight of 135.4 miles in 152 minutes, not counting stops for fuel at Poughkeepsie and Spuyten Duyvil, Curtiss put America back into the headlines. His achievement, though it fell short of Paulhan's in distance, demonstrated again that the quickest way to link two cities was by air. It also won him, by acclamation, the third leg on the *Scientific American* trophy, which entitled him to its permanent possession. The million spectators who applauded Curtiss's arrival in New York had barely time to catch their breaths when Charles K. Hamilton won a like amount of money from the *New York Times* and the Philadelphia *Ledger* by traveling, in another Curtiss plane, from New York to Philadelphia and back in a day—175 miles in 3 hours 27 minutes of flying time (with stops for refueling at Philadelphia and Perth Amboy, N.J.).

At the time Curtiss thrilled Americans with his record flight, showing "an unsuspected command of speed and distance," President William Howard Taft issued a statement to the press:

I am intensely interested in what Mr. Curtiss has done. It seems that the wonders of aviation will never cease. I would hesitate to say that the performance of Mr. Curtiss is an epoch because tomorrow we may hear that some man has flown from

New York to St. Louis. Mr. Wright told me at the time the ten mile flight from Fort Myer was made that the chief difficulty was flying over unknown territory. Mr. Curtiss seems to have surmounted this difficulty. His flight will live long in our memories as the greatest.

The flight did indeed live long in memory; but it was far from the greatest. So rapid was the progress of aviation that not much more than a year later the Albany trip would look like a short excursion—and Taft's vision of a flight between St. Louis and New York would be fully realized.

In Europe meanwhile, the racing tempo increased as the summer progressed. Delighted by the *Daily Mail*'s reporting of the London–Manchester race, its patriotism roused by the gallant effort of Grahame-White, the British public was ready for more meets and more competitions. King Edward VII's empire was the mightiest in history, holding sway over nearly one-fourth of the world's land area and including nearly one-fourth of the earth's inhabitants. It was inevitable that the idea of British supremacy in the air should be fostered by a newspaper and that interested promoters should capitalize on the theme.

At the end of June the city of Wolverhampton staged an all-British show. Participants included Cockburn, bearer of British colors in the Gordon Bennett race, who won the quick-start contest in the same Farman machine; Graham Gilmour, featured in numerous events with a Blériot; the Honorable Alan Boyle, who, piloting an Avis light monoplane, became the first to undertake a cross-country monoplane journey in England; Captain G. Dawes, one of the earliest English military pilots, flying a Blériot; and Lieutenant L. D. L. Gibbs, who flew a Henry Farman specially made for racing. Gibbs had previously suffered the loss of a machine in Spain, where a mob had become incensed because he would not fly in windy weather and had burned his plane. At Wolverhampton hard luck hit again: he ran into the "wash" of another machine and suffered a severe fall.

Early in July the pick of the pilots competed at Bournemouth, on the Channel. A race over the waves to the Needles and back was won by a Blériot piloted by the Frenchman Léon Morane. Grahame-White and one of his pupils, a nineteen-year-old Philadelphia expatriate named J. Armstrong Drexel, also completed the course. And Robert Loraine, a leading London actor who had taken up flying under the name "Mr. Jones," sent cheers through the crowd by speeding back to the mainland after being forced down overnight on the Isle of Wight by stormy weather. Loraine was not so lucky on September 11, however, when he flew 52 miles from Holyhead across the Irish Channel only to fall into the sea just short of his destination; he then proceeded to swim the last hundred yards to Irish soil at Howth Head. For his derring-do he received a silver medal from the Royal Aero Club of Great Britain.

The Bournemouth meet, unhappily, took the life of Charles S. Rolls

—the first English victim of an aeroplane crash. Rolls, founder of the Royal Aero Club, was on the way to making his name as famous in the air as it was on land. Blériot had taken the cross-Channel prize in 1909 before Rolls was ready; and Jacques de Lesseps, the twenty-seven-year-old grandson of the engineer Ferdinand de Lesseps (who built the Suez Canal), had deprived him of the honor of being second by winning the Ruinart champagne firm's prize of £500—to which had been added a £100 *Daily Mail* cup—on May 20. But on June 2, Rolls had achieved the distinction of becoming not only the first Britisher across the Channel but also the first aviator to complete a nonstop flight to the French coast and back.

On the second day of the Bournemouth meet—July 12—Rolls won the slow-speed contest at 25.3 m/hr. Then, flying one of the first French-built Wrights fitted with wheels instead of launching skids, with a new horizontal tail surface between the twin rudders designed to operate with the front elevator, he was coming in during the accuracy-of-landing contest when a sudden movement of his controls appeared to tear away the tail-piece. The whole rear assembly was sheared off by a propeller, and the machine plunged to earth from a height of 100 feet. Various causes were adduced for the accident—too sharp a descent, a gust of wind that strained the structure—but it is probable that Rolls sacrificed his life to avoid the crowd. Too low and too near the enclosures, he may well have realized the danger too late to maneuver in safety. A bronze statue of the aviator, in flying suit, cap, and leggings, bearing a scale model of the Wright biplane, was subsequently erected in Bournemouth.

Many of these flyers went on to Scotland to perform in a meet at Lanark, in the rolling hills near Glasgow; others returned to Blackpool, where a second contest had been organized for August. At the Scottish tourney, held between August 6 and 13, appeared Drexel, McArdle, and Gilmour; the Italian Bartolomeo Cattaneo; and James Radley, a coming speed champion—all with Blériots. G. Blondeau was there with a Henry Farman, as was Captain Bertram Dickson—who won the slow-speed contest at the uncanny rate of 21.29 m/hr. Grahame-White made the world's quickest getaway to date (also with a Henry Farman), in an incredible run of only 20 feet 9 inches.

Among the newcomers was sixteen-year-old Marcel Hanriot, the youngest licensed pilot in the world—whose father, René, had produced a promising new monoplane. Another was G. C. Colmore, flying a biplane built by the Short brothers—who were soon to market the first successful machine with two motors. Accidents were plentiful. G. P. Küller, the second flyer to be licensed in Holland, fell harmlessly into a wood when the motor of his Antoinette stopped abruptly.

But on the Continent, during a second meet held at Rheims, another Antoinette killed Charles Wachter—chief instructor of the school and son-in-law of Levavasseur—when both wings folded during a steep dive with the engine running. Wachter was the ninth to die in powered flight. It was obvious that the new sport was not exactly safe—and the public, while clamoring to watch, was not anxious itself to fly. But if the fillip of danger was ever present, it was no deterrent to steady progress.

The ability of pilots to find their way around the landscape had so improved that it was possible to think of an open race from one town to another. Such a contest was announced in France for June 6, 1910, as the culminating event of another "Grande Semaine d'Aviation." The course was between Angers and Saumur, in the department of Rochefort-sur-Loire; the distance was 43 km; and the entrants numbered nine.

The Angers–Saumur race was conceived by an early aeronaut and pioneer experimenter in powered flight, Robert Marie Jules René Gasnier (born 1874)—who with his brother Pierre formed an inventive fraternal team like the Wrights, Farmans, Voisins, and Telliers. Beginning in 1905, René Gasnier, an imposing man with a luxuriant black beard, had made balloon ascensions in various parts of Europe and the United States. Then, in 1908, the Gasnier brothers—like so many others fired by reports of heavier-than-air exploits—tested a homemade aeroplane. With René in the pilot's seat, Pierre lay in the grass to observe whether the wheels of their pusher biplane, employing an Antoinette 50-hp motor, left the ground.

Before his death in 1913, René not only staged the Angers–Saumur contest but two years later promoted the much more challenging Circuit of Anjou. Pierre, who became a pioneer civilian and military aviator, lives today at the family seat near Angers; in 1950 he was given the rank of commander in the Legion of Honor, for his services to French aviation. A

Robert Martinet prepares to take off in his Henry Farman biplane at the start of the first town-to-town race, between Angers and Saumur, France, June 6, 1910. (Pierre Gasnier du Fresne)

monument to René overlooks the picturesque site along the river Loire where his first trials were conducted.

At the start of the Angers–Saumur race, vagaries of the wind caused a long delay. More than two hundred thousand impatient spectators had gathered to witness the event. The Loire was jammed with boats, and the road with automobiles; pretty Red Cross nurses had taken up strategic stations to render first aid to anyone tumbling out of the blue. The trinational entry list of pilots and their machines read as follows: Emile Aubrun (Blériot), André Crochon (Henry Farman), Captain Bertram Dickson (Henry Farman), Georges Legagneux (Sommer), Robert Martinet (Henry Farman), the German Walter de Mumm (Antoinette), and Marcel Paillette (Sommer). Most of these flyers had received their pilot's certificates only recently; their temerity in signing up for such a risky excursion led to cries of protest from the local press.

When the starting flag was finally lowered, only three planes rose into the air—the Sommer biplane of Legagneux and the Farmans of Martinet and of Dickson. All three machines were powered by the Gnôme 50-hp motor, so that odds were even save for luck and skill. At Saumur, where half the crowd was waiting, cheers turned to delirium when the first dot to appear in the distance was followed shortly by a second. Martinet, the winner, had covered the distance in 31 minutes 35 seconds, Legagneux in 36 minutes 45 seconds. Both landed adroitly near the grandstand and were carried off shoulder high, while the band blared forth the "Marseillaise." Dickson landed on a flooded portion of the field—into which it seemed he would have to jump from his lofty seat in the Farman. However, in a touch of the comic, he was rescued by a member of the Horse Guards, who brought his mount alongside a wing and bore off the marooned aviator dry shod.

Previous doubts forgotten, the Paris papers joined in the paean of praise next day. "For the first time in the history of aviation a number of aeroplanes engaged in a simultaneous cross country race . . . ," jubilantly chronicled the *Herald*. "A new era . . . of triumph" was the *Petit Parisien*'s salute to progress. The race was to prove, said *Le Gaulois*, ". . . a definite forward step in the unceasing advance of the new sport."

With the stimulus of the Angers–Saumur success, and determined not to be outdone by the *Daily Mail* and its London–Manchester contest, the French newspaper *Le Matin* sponsored in August a ten-day aerial tour out of Paris, starting and ending at Issy-les-Moulineaux with controlled stops at Troyes, Nancy, Mézières, Douai, and Amiens. Uncounted thousands of Parisians watched the start of the Circuit de l'Est, as it was designated—the world's first long race across open country. From stop to stop the newspaper accounts poured in—and *Le Matin*'s circulation soared like the aeroplanes.

To a main prize of 20,000 francs were added special prizes for local exhibitions and contests. The result was a decisive victory for the house of Blériot and its characteristic workhorse, the Gnôme engine. Of eight starters only two were able to overcome the mechanical difficulties that beset the field—and those two flew the winning combination. Alfred Leblanc, steadfast friend of Blériot, was first with a total time for the course of 12 hours 1 minute 1 second; his teammate, Emile Aubrun, covered the 785 km in 13 hours 31 minutes 9 seconds.

The Blériot aeroplane was, literally, in the ascendant everywhere. All that summer the altitude record was steadily pushed upward by aviators flying Blériots. At Blackpool on August 3, Georges Chavez (born in Paris in 1887 of Peruvian parents) reached 5850 feet. Chavez, an engineering student and noted soccer player, had entered the Henry Farman school in December 1909 and soloed on February 10, 1910; he then switched to the Blériot, with which he soon won prizes at Biarritz and Tours. Drexel raised the ante at Lanark on August 11 to 6605 feet; on August 29 the ceiling was lifted again by Morane at Le Havre to 7054 feet, and then on September 3 at Deauville to 8471 feet. Chavez returned to the fray at Issy on September 8, putting the record up another notch to 8487 feet. Looking back a year to the 508 feet credited to Hubert Latham at Rheims on August 29, 1909, the advance had been phenomenal.

So successfully was man reaching for the skies that the most ambitious projects now seemed logical as well as feasible. As the curtain raiser for an air show at Milan, the committee in charge announced *la traversata delle Alpi in aeroplano*—nothing less than the crossing of the Alps by aeroplane. Italy's incipient interest in aviation had been stimulated by a meet at Verona from May 22 to May 29, with prizes totaling 200,000 lire (although 40,000 lire of this amount was not awarded, for lack of entries in an international speed contest). Again the names of the flyers were familiar: Paulhan, Efimoff, Duray, Chavez, Molon, Küller, Cattaneo—the last named being the only Italian. Three types of plane, all French, were seen: the Blériot, the Voisin, and the Antoinette. As a sporting proposition the Verona meet was considered a success, but financially it was a failure. Organizers of the fall meet at Milan, set to run from September 25 through October 20, hoped for a much higher attendance by featuring, during the preceding week, the first crossing of the Alpine heights. In deference to this plan, the president of the Società Aviazione Torino agreed to cancel his contract for an exhibition by French flyers at the Mirafiori field near Turin, scheduled for the first fortnight in September.

The perils of the mountain passage were self-evident. It would not be the first time that the Alps had been crossed by air—for the French aeronaut Francisque Arban had ballooned his way from Marseilles to Turin on September 2, 1849, rising to 4000 meters in a trip that lasted all night. But traveling in the flimsy, underpowered machines of 1910 was another matter. From the sleepy Swiss village of Brigue, at an elevation of 900 meters, the contestants would have to tackle the Simplon Pass—requiring an altitude of at least 2100 meters. If a plane's motor was found to be safely functioning in the cold, thin air, there was still the capricious wind—violent and unpredictable at that time of year. One of Italy's few aeronautical engineers, Canovetti, struck a note of pessimism, pointing out the grave responsibility resting on those who had organized such an event and suggesting that too little time had been allowed for preparation. But dreams of glory had gone to the collective head of the Società Italiana di Aviazione Milanese. There could not be a finer sporting gesture, one of the members wrote in the periodical *Aviatore;* it was "audacious and patriotic; if it succeeds, it will bring Italy into the front ranks of nations concerned with flying . . . [it] would show more than anything" that Italy was the equal of the United States, France, and England.

Despite the dangers, thirteen aviators registered for the event. Once the Simplon had been surmounted, the wide valley of Bognanco and the town of Domodossola would lie ahead; then the borders of Lake Maggiore, the gentle Lombardy plain, and the beckoning city of Milan—a total distance of 75 miles. The scheme had the full support of Italy's National Commission of Aerial Touring, the Italian Touring Club, the Società Immobiliare Lombardo-Veneta (the commercial sponsor of the race), the Swiss organizing committee at Brigue, and local committees at the way stations of Domodossola, Stresa, Pallanza, and Varese. Prizes totaled 100,000 lire, of which 70,000 lire was for the crossing alone. It looked like easy money—once the heights were scaled.

Nevertheless, conscious of the frailties of both men and machines, the committee decided to limit the competition "to those aviators who are admitted by the sporting commission . . . which may refuse an inscription without revealing the motive." Under this rule only five of the applicants were adjudged to be qualified to compete: Chavez, an inveterate high flyer, with his Blériot; Charles T. Weymann, an expatriate American sportsman, who was soon to be the leading exponent of a new, enclosed-fuselage monoplane—the Nieuport; Eugen Wiencziers, the German pilot of an Antoinette; Cattaneo, with a Blériot; and Marcel Paillette, with his Sommer. When the last three of these withdrew, the game became a two-sided one between Chavez and Weymann.

September 18 fell on a Swiss holiday, and—to the annoyance of the waiting aviators—no flights were permitted. Thus the tail end of a spell of perfect weather was lost. Chavez made the first try on the 19th, but descended quickly with the remark that flying machines were but "playthings of the winds in the passes." The young Peruvian, so the rumor went, was romantically involved with a charming Parisienne, for whose favor he had determined to fly higher than anyone else—like a knight in a medieval tourney entering the lists against all would-be champions. Weymann too turned back; and unfavorable conditions continued to prevent flying on the 20th. On the 21st Weymann tried three departures but was unable to gain altitude; his carburetor covered with frost, he was forced to admit defeat.

Monument at Brigue, Switzerland, to Peruvian flyer Georges Chavez, whose plane plummeted to a fatal crash after the first successful crossing of the Alps. Chavez's dying words— "Higher, ever higher"—became the motto of the Peruvian air force.

Finally, on the 23rd—the day before the Alpine contest was to close —Chavez succeeded in taking off from Brigue's postage-stamp airfield at 1:29 in the afternoon. He flew steadily through the turbulent air and, to the indescribable excitement of those who watched, appeared over Domodossola 42 minutes later. He then prepared to bring down his frail craft. It is still uncertain whether a wing of the machine, weakened by vicious gusts, gave way or whether Chavez was so benumbed by the cold in the open cockpit that he could not control his final descent; but at the last moment he crashed from a height of 10 meters, breaking both legs. Awarded half the prize money for his effort, Chavez died in the Domodossola hospital two days later, still semiconscious, murmuring, "Higher, ever higher." These famous last words—*Arriba, Más Arriba*—became the motto of the Peruvian air force; and September 23 is National Aviation Day in Peru.

The pioneer passage that Chavez had set out to forge was not to be completed for three years—when Jean Bielovucic, another Peruvian, resolved to vanquish the mountain fastness in the name of his unfortunate countryman. A bicycle rider before he took up flying, "Bielo" not only was a strong contender in many European races but helped to introduce the aeroplane in Peru. Noting that the Alpine winds, with their treacherous eddies, blew less often in winter, he carefully planned his attempt for the coldest season of the year. On January 25, 1913, after waiting two weeks for the right moment, he covered the same route that Chavez had attempted, in an Hanriot monoplane without incident; it took him exactly 26 minutes. Bielovucic fought as a volunteer in the French air force during World War I and became a chevalier of the Legion of Honor.

Although he did not live to taste the fruits of victory, Chavez received worldwide recognition for his exploit. The wreckage of his plane was restored for permanent display in the Military Historical Museum of Peru; and, fifty years after the event, a monument and fountain, with a bas-relief of the flyer's head and shoulders by the Swiss sculptor Vuilleumier, was dedicated in his honor at Brigue. Another marker stands where he crashed at Domodossola—while a monument in Lima, to which his remains were brought in 1957, keeps his name alive in the homeland he never saw.

The Chavez tragedy cast a shadow over the Milan meet. Recriminations poured forth in the press. What did such a reckless exhibition in the mountains prove, anyway? Merely that the human race possessed great courage, said the critics bitterly; nothing whatever of scientific value was gained. Aviation had a nobler purpose than to create heroes and martyrs. At that stage of its development, it was inexcusable that such unnecessary risks be taken.

The rebuttal lay in the crowds that flocked to Milan's Taliedo aerodrome, outside the Porta Vittoria, where forty-five aviators—undeterred by the accident and lured by 320,000 lire in prize money—had signed up. Only twenty-seven of these actually participated. Foreign planes with foreign motors, and foreign flyers licensed in foreign schools, predominated; and nearly all were French.

Italian prestige, by contrast, was not high. Planes built by the Marquese Filiasi, by Ferro and Balbi, by the Antoni brothers, were seen but not heard from; nor was the Bacciega biplane or the Marra-Altieri (first powered by a Miller motor, then by an Anzani) able to give an account of itself. Even the tested triplane of a dedicated engineer like Aristide Faccioli was withdrawn in the face of the French competition. At that period there was in fact only one promising heavier-than-air craft in Italy: the Asteria biplane, derived from the machines of the Wrights and of Henry Farman that had flown at Centocelle, put together by the engineer Darbesio, and piloted by the Piedmontese aviator Emilio Pensuti and by Lieutenant Giulio Gavotti of the army. It would be several years more before such great aeroplanes as those of Gianni Caproni, the World War I designer, took their rightful place in Italy's blue skies.

The Milan meet was marred by the first collision to take place in the air. Captain Bertram Dickson, who had headed the list of prize winners at Rouen in June and had won all the prizes in a meet at Tours, was flying his slow Farman over the aerodrome at a height of 40 meters when the Frenchman René Thomas rammed him in the rear with his speedier Antoinette. Both machines were wrecked, and both pilots badly hurt. While there were no immediately fatal consequences, the mishap to Dickson formed the basis of a legend that grew up around him. A short time before at the British Museum, he and his sister had inspected the mummy of the priestess of Amen Re, said to bring ill luck to anyone who came in contact with it. Within six weeks Miss Dickson had been shipwrecked on the Albanian coast. Dickson himself, lying injured in the hospital, learned that his bank had failed and that he had been brought to the edge of ruin. A court of inquiry held him responsible for the collision because he had been making an exhibition flight, whereas Thomas had been taking part in a competition; and he was ordered to pay £600 in damages. Dickson did not fly again until he became associated with the Bristol Aeroplane Company in March 1912; but on September 28, 1913, he died, having never fully recovered from the severe shock of his injuries.

Not all the prizes to be plucked out of the air were limited to Europe. The flying fever had broken out in Boston with a gathering of English and American aviators, organized by the Harvard Aeronautical Society for the period September 3 to 13, and a total of $100,000 was contributed for the winning flyers. Among the purses offered, the Boston *Globe*'s donation of $10,000 for a race around Boston Light, a distance of 33 miles over water, was the main attraction.

The field at nearby Squantum displayed a cross section of America. Open touring cars jammed the parking space; scores of Social Register celebrities attended; prominent leaders from Boston, New York, and Washington—including the substantial bulk of President Taft—were seen scanning the heavens; and thousands of other enthusiasts packed the bleachers. Flying for England, the dashing Grahame-White was the man of the moment, mingling convivially with American sportsmen and sportswomen. The British pilot graciously acknowledged the applause that rose when he won the *Globe* prize with his Blériot in 34 minutes 11⅕ seconds, and in addition a second prize for speed: 5¼ miles in 6 minutes.

Grahame-White's teammates were Thomas Octave Murdock (Tom) Sopwith and Alliott Verdon Roe—both soon to develop into front-rank designer-constructors and both eventually to be knighted for their contributions to aeronautics. Sopwith, well off financially, had drifted into aviation from yachting; he had learned to fly in a Howard-Wright biplane. In December, Sopwith was to win the Baron de Forest prize of £4000 for the longest flight from England to the Continent in a British-built machine. Later still, he would be a constructor in his own right, producing in World War I a fighter plane that became a favorite of the Allies—the Sopwith Camel. Roe, on the other hand, was one of the earliest experimenters in England. He began at the beginning, first winning a *Daily Mail* competition for model aeroplanes, then achieving first flight at Brooklands on June 8, 1908, with a shaky Antoinette-powered biplane of his own construction—the Roe I. Shortly afterward Roe built a triplane with his brother Humphrey, which formed the basis for the highly successful AVRO Company. At the Harvard meet, however, he ran into bad

Pioneers of Flight (biplanes)

Odier-Vendôme

Maurice Farman

Voisin

Henry Farman (military)

Voisin Rigal

Caudron

Henry Farman

Sommer

Blériot

Morane

Blériot (military)

Morane-Saulnier

Antoinette

Nieuport

Sommer

Deperdussin

luck with his triplane and achieved only a few seconds of flight before coming to grief.

Americans on the entry list included Glenn Curtiss; Walter Richard Brookins of Dayton, the first American pupil of the Wrights, who had promised him a plane of his own while he was still a boy; Ralph John-stone, a former vaudeville performer whose stunting with the Wright machine was soon to cost him his life; Charles F. Willard, reputed origi-nator of the expression "holes in the air," used to describe downdrafts; Earle L. Ovington, an exhibition flyer from Garden City, Long Island, with a Queen monoplane; and Clifford Burke Harmon, who on August 20 had won the $2000 *Country Life in America* trophy by flying a Henry Farman plane from Garden City across Long Island Sound to Greenwich, Connecticut—25 miles in 29 minutes. Harmon, a subsequent president of the Aero Club of America, distinguished himself at the meet by taking the Harvard cup for the best amateur bomb dropper, as well as prizes for speed, duration, and slow time. His name in aviation was perpetuated by his donation of the annually awarded Clifford B. Harmon trophy.

Throughout the Harvard meet the weather was ideal. For the statisti-cally inclined it may be noted that a total of 29 hours 27 minutes 7⅖ seconds was spent in the air and that an overall distance of 631 miles 3617 feet was flown (not counting exhibitions and trials). No fewer than 170 "bombs" were dropped, "without fatal injuries to anyone," as a con-temporary news account stated, "although the occupants of one auto had their feelings badly hurt when an egg-bomb struck their car and splashed them." The altitude contest was won by Brookins at 4732 feet—higher than the world mark of 4380 feet he had established at Indianapolis on June 14 but substantially lower than the new record of 6259 feet he had set at Atlantic City on August 7.

On the agenda for October was an international aviation meet that surpassed anything yet seen in the United States. From the 22nd to the 30th of that month, New York played host to twenty-seven of the world's leading flyers in a memorable three-nation tournament held at the Belmont Park racetrack on Long Island. Momentarily at least, the pendulum of activity seemed to be swinging back to America.

A total of $72,300 in prize money was at stake—of which $10,000 was earmarked for the winner of a race around the Statue of Liberty. Another $1,000 was thoughtfully set aside "to be distributed among the mechanics of the aviators as a recognition of their services." Top billing, however, was given to the second contest for the Gordon Bennett trophy—and the first on American soil. Patriotic citizens turned out in force to watch the preliminaries, confident of seeing the Star-Spangled Banner waving victo-riously at the finish.

American hopes were pinned on an all-star Wright delegation: Walter Brookins, selected to pilot the dark-horse Model R racer, of Orville's creation—a junior-size "headless" machine, lacking a front elevator and with a 21-foot wing span and eight-cylinder, 60-hp engine; Ralph Johnstone, a specialist in high flying; his rival in altitude, Arch Hoxsey (who had taken up former President Theodore Roosevelt at St. Louis on October 11); and either Phil O. Parmelee or J. Clifford Turpin. In the Curtiss camp were Charles F. Willard; James C. (Bud) Mars, a former balloonist and parachute jumper; J. A. D. McCurdy, charter member of the Aerial Experiment Association; and Eugene Burton Ely, whose later exploits were to earn him a place in America's Aviation Hall of Fame. Charles K. Hamilton entered his Hamiltonian—a biplane modeled on the Curtiss and powered by an eight-cylinder, 110-hp motorcar engine designed by Walter Christie. Captain Thomas Baldwin appeared with the Red Devil, constructed on the same order; and Clifford Harmon cast his lot with a Henry Farman. Flying Blériots were John B. Moisant, who in August had made the front pages by carrying his mechanic and a yowling kitten, Mademoiselle Paree, from Paris to London in just under three weeks; and J. Armstrong Drexel, fresh from European successes. Todd Shriver, Curtiss's former mechanic, was named pilot of a confection known as the Howard-Dietz biplane. And millionaire Harry S. Harkness, who had taken his license just three days before the meet, was enjoying the distinction of being the only man in America to own an Antoinette.

France entered Alfred Leblanc, Count Jacques de Lesseps, Emile Aubrun, René Simon, and René Barrier, all flying Blériots; Hubert Latham, who was to captivate New York with his handling of the Antoinette; and Edmond Audemars and Roland Garros, two exponents of the low-slung, low-powered Demoiselle—both of whom had learned by trial and error at Issy-les-Moulineaux and who could be expected to gain little but experience from the meet. René Thomas with his Antoinette had been listed to compete, but was prevented by the aerial collision at Milan.

England presented her favorite flyer, Claude Grahame-White. After the Harvard events he had paid his respects to Washington officialdom, startling the capital on October 14 by nonchalantly making a landing on and then a takeoff from West Executive Avenue, a narrow thoroughfare between the White House and the State, War, and Navy Building. Grahame-White brought both a Henry Farman and a Blériot to New York. With James Radley and William McArdle in Blériots and Alec Ogilvie operating a Wright to complete the roster, the British team had the fewest entrants in the meet.

The affair at Belmont, like its Rheims counterpart, was "the" social and sporting event of the decade. Anyone who was anybody felt obliged to attend—conveyed to the scene by special trains of the Long Island Railroad, by trolley, or in the latest-model touring car. Society's "Four Hundred" were out in force, to see and be seen. Vanderbilts, Whitneys, Goulds, Drexels, Hitchcocks, and Havemeyers filled the boxes, the paddock, or the grandstand, while thousands of ordinary mortals paid a dollar apiece for general admission. Clarence H. Mackay—financier, art patron, and president of the Postal Telegraph Company; Harold F. McCormick, of the famed Chicago machinery dynasty; Lieutenant Governor Timothy L. Woodruff of New York; and the prominent Boston sportswoman Eleonora Sears were the cynosures of all eyes. Entire pages of the metropolitan press were given over to the doings at the track, where horses and jockeys had been displaced by a new order of winged steeds. Columns of newsprint described the tailored costumes, the long velvet dresses, and the modish beaver hats of the ladies, freely interspersed with the technical details of an aeroplane's construction. Flying buffs glibly discussed the merits of monoplanes versus biplanes: "In Buenos Aires," observed an Argentine visitor, "we prefer the monoplane—the biplane is a woman's machine." But the women raved about the lines of the glamorous Antoinette and fought for introductions to the aviators as if they were gods on the slopes of Olympus. And as if the aerial "daily double" at Belmont were not dramatic enough, Alan R. Hawley and Augustus Post had just disappeared over the Canadian wilds while competing in the annual Gordon Bennett balloon race. This report, coming on top of the news of Walter Wellman's quixotic attempt to cross the Atlantic in a dirigible, set tongues wagging in a lighter-than-air counterpoint to the hum of powered machines over the flat Long Island landscape.

The weather on Saturday, October 22, could scarcely have been less auspicious to introduce New York to the wonders of flight. When the smoke bomb went off at 1:30 P.M., a cold fog was hanging low over the field, mixed with a steady drizzle that played havoc with ignition systems. Ulsters, furs, felt hats, and umbrellas were much in evidence; attendance figures showed a scant two thousand. Rather than disappoint the faithful, however, a handful of intrepid pilots went aloft. In wary hops that never took them far from the ground, they proved at least that they could navigate in a rainstorm. A height of only 742 feet was enough to win the altitude contest on that inaugural day; the prize was taken by Arch Hoxsey.

Sunday, the 23rd, was no more propitious for instilling confidence in the capabilities of a flying machine. A "tempest" of 20 to 30 m/hr kept entrants herded in the hangars—till Grahame-White took his Farman into the bumpy air. Ten thousand onlookers held their breath while the craft pitched and tossed around the track like a catboat in a rough sea. On landing, the machine was wrenched out of the aviator's control and badly damaged. Undaunted by this prospect, Moisant ordered his monoplane prepared for flight; but it was rudely blown out of the hands of seven struggling men before he could even clamber in. With two wrecks on its conscience, an apprehensive committee forbade further attempts.

In the days that followed, there were other inclement periods, when the blue "No Flight" pennant stood stiff in the breeze, as well as lengthy delays when cantankerous engines kept mechanics in endless frustration. During such interludes the crowd listened patiently as the Seventh Regiment Band played such popular tunes of the day as "Every Little Movement Has a Meaning All Its Own." As the meet wore on, those who came to watch the "airships" (as the program had it) learned that there was less wind at twilight; and many were content to wait all day for their reward.

At other times, when white "Flight Probable" or red "Flight in Progress" flags hung motionless and sparkling autumn sunshine warmed the scene, continuous activity made up for previous disappointments. Contrast with the familiar Curtiss and Wright biplanes was provided by the array of monoplanes from Europe. From the twenty hangars and four huge tents sorties were made, for the first time on the Eastern Seaboard, by the latest-model Blériot, Antoinette, and Demoiselle ("The Infuriated Grasshopper," albeit no contender in speed, altitude, or duration, was angrily determined to be heard).

On October 24 a total of ten machines in the air at the same time touched off a storm of applause. On the 25th, Drexel—conspicuous in crash helmet—set a new American altitude record of 7185 feet; however, he held it only till the next day, when Johnstone, in the teeth of a snow flurry, climbed 228 feet higher. (On October 31, Johnstone outdid himself by toppling the world altitude record with an ascent of 9712 feet—or 2960 meters—more than twice the height mark set by Brookins at Indianapolis only a few months earlier.) De Lesseps became lost in the clouds; and alarm was piled on anxiety when Latham's motor suddenly went dead at 3000 feet and he was forced into an interminable volplane.

On the 26th a dozen monoplanes and biplanes flitted through the gloaming, motors purring like gigantic sewing machines, pilots waving to the grandstand in the homestretch. Drexel, de Lesseps, Parmalee, Brookins, Ely, Mars, Simon, Grahame-White, Moisant, Radley, Latham, Audemars, Barrier—the crowd greeted each with enthusiasm as they passed by like the familiar figures on a carousel. On the 27th, Hoxsey and Johnstone—"The Heavenly Twins"—during one of their classic altitude duels, found themselves traveling backward, at the mercy of a wind stronger than the speed of their machines.

With the tournament a week old, the Gordon Bennett race was run on October 29. Since Rheims, speeds had increased appreciably, and a possible mile a minute was now predicted. Adverse weather conditions had prevented regular elimination trials for the defending American team; but on the basis of performance and reputation, those named to represent the United States were Brookins, Drexel, and Hamilton, with Mars, Moisant, and Hoxsey as substitutes. The official French team was composed of Aubrun, Latham, and Leblanc. Britain was represented by its three-man list at the meet—Grahame-White, Ogilvie, and Radley.

There were six starters, two from each country. America's high expectations had been dashed when the Wright Model R (known as "The Baby"), piloted by the twenty-two-year-old Brookins, was eliminated before the race by an unexplained fall from 50 feet. Disappointment was mixed with relief that Brookins was unhurt; but only crumpled wreckage remained of what had been considered the fastest craft in the United States.

Over the 100-km (62.14 miles) course the flyers ground out lap after lap in pursuit of lasting renown—and the first prize of $5000. Shaking under the vibrations of its 100-hp motor—the most powerful Gnôme yet built—Grahame-White's Blériot quickly set the pace, reaching the finish in 1 hour 1 minute 4¾ seconds. Not even the magician Harry Houdini, who had added to his bag of tricks that year by learning to fly a Voisin-type biplane, could have pulled an American victory out of the hat in the face of that performance. Grahame-White easily captured the cup for the Royal Aero Club of Great Britain, with an average speed of 61.3 m/hr.

In a machine identical in every respect, Leblanc strove desperately to nose out his English rival. Rounding the pylons in vertical banks that left the onlookers gasping, the Frenchman stepped up his plane to an estimated 70 m/hr. In vain: after covering 95 km in 52 minutes 9 seconds, his gas line broke on the last lap, and the Blériot hurtled into a telegraph pole. Why the pilot should have escaped with a couple of minor head wounds, and the plane with a splintered propeller and moderately damaged chassis, is one of those miracles in which the early annals of aviation abound.

Second place was saved for America by Moisant, with his Blériot rated at half the winner's horsepower; his time was 1 hour 57 minutes 44 seconds, including a stop of 38 minutes. Drexel, in still another Blériot, was compelled to quit after flying 35 km in 26 minutes 4 seconds. Hamilton was not able even to start.

Two others finished the grueling course. It took Alec Ogilvie 2 hours 6 minutes 36 seconds—more than twice the time of the victor—to gain third place for England in his 30-hp British-built Wright. Latham was an inglorious fourth, spending 5 hours 47 minutes 53 seconds in alternate hops and repairs. (The gorgeous Antoinette, however, left a vivid impression on New York: a large toy facsimile, covered in yellow silk and powered by rubber bands, was a best seller at F. A. O. Schwartz that Christmas.) Sitting high on the slender fuselage and manipulating the large control wheels on either side, Latham evoked cries of admiration as he thundered past, jauntily saluting the mass of upturned faces. Later, during a small meet at Halethorpe in the suburbs of Baltimore, he was to reach another pinnacle of fame by soaring over the crowded streets of the city—in order that an invalid millionaire, R. Winans, might see what a plane looked like in flight. Latham was awarded $5000, through the offices of the Baltimore *Sun,* for his demonstration.

The Gordon Bennett was a contest to be long remembered by Americans, who gaped at what the foreign machines could do. "In the decades that lie hidden in the future," rhapsodized a *New York Herald* reporter, "there may be possibilities that may make appear slight and trivial the incidents of yesterday, but to those who are now witnessing the events of aviation pioneering, the day on which the international trophy was won for England by Mr. Claude Grahame-White must be regarded as writing upon the pages of history an advance in science such as never before has been recorded."

Still another climax, however, was to come on the next-to-last day of the tournament. A hundred thousand persons had gathered on Sunday, October 30, to watch the race around the Statue of Liberty. A flight of 34 miles across land and water was enough to tax even the most skilled and courageous; but "the dangers of flying above a city like Brooklyn," as one scribe put it, "are regarded by aviators as almost beyond the bounds of recklessness. A fall means almost certain death."

Nevertheless, the victor of the Gordon Bennett classic, Grahame-White, set off to negotiate the perilous course—to the cheers of thousands on the city housetops—for what appeared to be a certain triumph. De Lesseps was a good prospect for second place. With the Curtiss and Wright camps depressingly unprepared for this final test, America's chances looked dim indeed. And minutes later they appeared to fade entirely as Moisant, hurriedly taxiing to beat the deadline at the starting point, crashed into Clifford Harmon's parked biplane, hopelessly entangling the two machines. But the impulsive Moisant was not yet ready to quit. Bidding desperately for a chance to retrieve the national honor, he borrowed $10,000 from his brother Alfred—a gamble on the chance of winning the prize, of the same amount. Rushing to a telephone, he importuned the slightly injured Leblanc to sell him a brand new Blériot lying idle in the Frenchman's hangar. Leblanc's sporting blood (or perhaps the realization that the cost of his machine had been only half the offer) came to the fore: he called it a deal. No matter that Moisant had never seen the Blériot before; without even tuning up he drove it down the field like a scared jackrabbit and took to the air.

The epic flight that followed would never be forgotten by those who ticked off the minutes and seconds, straining for the first glimpse of the returning flyer. People spoke in hushed tones, as if the destiny of the nation depended on the lone craft buzzing across New York Bay. Then, after half an hour, a black speck appeared on the horizon. When it was disclosed that Grahame-White's time had been beaten—by less than a minute—pandemonium swept the grandstand. In paroxysms of joy people threw themselves into each other's arms and howled themselves hoarse. Hats, canes, umbrellas, handkerchiefs were hurled into the air. Moisant was carried to the judges' stand on the shoulders of his friends—the hero of the hour.

It might be true that the United States had lost the meet itself, the Gordon Bennett race, and many of the lesser events. But the Statue of Liberty race had been won by an American—in as soul-satisfying a finish as the most rabid victim of the aviation craze could wish.

The celebration proved to be premature. Backed by the French flyers, the British team protested Moisant's win on the premise that he had started after the stipulated hour of 4 P.M. and should, therefore, have been disqualified. It took two years for the Fédération Aéronautique Internationale to decide the case in favor of Grahame-White—who was thereupon awarded the $10,000 prize, together with accrued interest of $600. This incident was one of the rare instances of ruffled feathers among the early birds.

After Belmont Park, Grahame-White returned to England and bought the weed-grown property at Hendon, which he and associates proceeded to develop into the biggest British flying field of its day. With unconscious infelicity, posters in London's Underground solicited public attendance as follows: "Flying At Hendon—The Quickest Way To Reach The Ground." But it was an investment that paid off: Grahame-White was reported to have later sold the aerodrome to the British government for half a million pounds.

The competitors at Belmont earned varying sums, as shown by an unofficial tabulation published after the meet. Moisant (if one includes the disputed Statue of Liberty prize) was high scorer, with $15,800. Other earnings were as follows: Grahame-White, $9,700; Johnstone, $6,625; Hoxsey, $3,675; Latham, $3,250; Aubrun, $1,250; McCurdy, $1,100; Radley, $1,050; Drexel, $800; Simon, $750; de Lesseps, $700; Brookins, $150; Barrier, $100; Mars, $100; and Willard, $50. Whether those who risked their necks in wind or rain considered the compensation adequate is unknown; but those who paid admission on days that flight took place undoubtedly got their money's worth. Like the meet at Rheims, the Belmont tourney ended, miraculously, without a fatal accident. Within three weeks, however, death came to one of the top performers—Ralph Johnstone—in a crash at Denver.

The United States was slower than France in issuing pilot's licenses and establishing the rules under which they could be granted. Regulations published in 1910 stated: "All candidates shall satisfy the officials of the Aero Club of America of their ability to fly at least five hundred yards, and of their capability of making a gliding descent with the engine stopped, before their applications can be entertained." Up to the time of Belmont Park, twenty-five such licenses had been issued. The first dozen of these were as follows:

No.	Name	Machine
1	Glenn H. Curtiss	Curtiss biplane and motor
2	Frank P. Lahm	Wright biplane and motor
3	Louis Paulhan	H. Farman biplane, Gnôme motor
4	Orville Wright	Wright biplane and motor
5	Wilbur Wright	Wright biplane and motor

6	Clifford B. Harmon	H. Farman biplane, Gnôme motor
7	Thomas S. Baldwin	Curtiss biplane and motor
8	J. Armstrong Drexel	Blériot monoplane, Gnôme motor
9	Todd Shriver	Curtiss biplane and motor
10	Charles F. Willard	Curtiss biplane and motor
11	J. C. Mars	Curtiss biplane and motor
12	Charles K. Hamilton	Curtiss biplane and motor

With the exception of the certificates given to Harmon, Shriver, and Mars—dated, respectively, May 21, September 17, and August 26, 1910 —no mention was made of the dates on which these first licenses were granted. It is, therefore, unclear why Curtiss rather than the Wrights was awarded license No. 1, or on what basis the others were assigned the order in which their names appear. As to the first question, the explanation may lie in the fact that the Curtiss flight of July 4, 1908, was the first to be officially witnessed by deputies of the Aero Club of America. As to the second question, it is probable that the first few licenses were awarded on "evidence" of an aviator's proficiency presented at Aero Club meetings, rather than on the basis of actual field tests. This supposition is borne out by the minutes of a meeting held on March 5, 1910, at which time it was simply moved and seconded that Paulhan be granted license No. 3. Beginning with license No. 13, all dates were published.

A separate category was established in August 1912 for flyers who qualified for an "expert aviator's certificate" under rules established by the Aero Club. Twenty-four such licenses had been awarded up to 1914. The first ten of these went to the following:

No.	Name
1	Max T. Lillie
2	Glenn L. Martin
3	Lieutenant T. DeWitt Milling, USA
4	Lieutenant Henry H. Arnold, USA
5	Captain Charles deF. Chandler, USA
6	Captain Paul W. Beck, USA
7	Lieutenant B. D. Foulois, USA
8	DeLloyd Thompson
9	Lieutenant Harold Geiger, USA
10	Lieutenant L. E. Goodier, Jr., USA

CHAPTER SIX

1911: *Year of the Great Races*

While aviation inundated the European countryside in a ground swell of popular activity, flying in the United States was all but confined to a few hardy demonstration performers, who carried it to the people at county fairs or carnivals. The first serious attempt to promote such local exhibitions was a direct outgrowth of the Belmont Park tournament. A group of contestants at loose ends after the meet were recruited by John B. Moisant—that restless and indefatigible soldier of fortune—for a tour of indefinite duration. This enterprise was to signal the birth of barnstorming. In a country where "Barnum and Bailey" was a household term, the flying circus was a typically American phenomenon. Traveling by special train with built-in repair shop, the entourage included a tent and a dozen roustabouts, a dozen ticket sellers and press agents, a dozen aeroplanes and their mechanics, and eight death-defying aviators.

Under the name Moisant's International Aviators, Ltd., the itinerant troupe started out from New York, opening in Richmond. It then went on to Chattanooga and Memphis, Tupelo (Mississippi), New Orleans, Dallas and Fort Worth, and Oklahoma City; back to Texas, with shows at Waco, Temple, Houston, San Antonio, and El Paso; across the border to Monterrey, Mexico City, and Vera Cruz in Mexico; and finally to Havana, Cuba. As might be expected, the odyssey was fraught with countless adventures and accidents—including a crash that killed the leader of the expedition himself.

The name Moisant (preserved today in the appellation for New Orleans's international airport) was a byword for the sensational. Claiming Chicago as a birthplace, though of French-Canadian origin, "John B." was a short, alert, amiable man with a round, bald head; a tanned, clean-shaven face; and scintillating black eyes. He had an ageless look, but was probably about forty at the time of Belmont; he spoke French and Spanish fluently and sported a diamond stickpin as well as a diamond ring. In 1910 the flamboyant John B. joined the group of novices who had established themselves in the Clément-Bayard dirigible hangar at Issy-les-Moulineaux; there he experimented, unsuccessfully, with monoplanes of his own design. One, a black-painted affair known as Le Corbeau ("The Crow"), had a perverse tendency to turn turtle. His career in the air really began in the spring of that year, when he purchased a two-seater Blériot, gave it a tryout, and without further instruction took his license next day at Etampes. On his third flight Moisant pulled off a grandstand play—landing at Issy with a passenger, Roland Garros, a few minutes before the starting gun of the Circuit de l'Est.

Moisant's International Aviators had a cosmopolitan cast of characters. There were a trio of Frenchmen: Roland Garros (who had been more an understudy than a star at Belmont), a sympathetic type who played the lead for Moisant and was billed as "The Cloud Kisser"; René Simon, whom the press agents called "The Fool Flyer"; and René Barrier—touted (with more imagination than realism) as "Record Holder of Flights over Cities." Then there was Edmond Audemars, of Geneva, Switzerland, whose short stature, toothbrush moustache, and bow tie inevitably earned him the name "Tiny"—pilot of "the smallest and most dangerous aeroplane in the world." Charles ("C. K.") Hamilton, a swashbuckling national hero since his Philadelphia journey, was fond of boasting that he had broken every bone in his body. John J. Frisbie, a whiskey-drinking Irishman older than the rest, had been an early Curtiss pilot; he was to be killed a year later at Norton, Kansas, having been goaded to fly in unsuitable weather by the jeers of a hostile crowd. And lastly there was Joe Seymour, whose flying ability (if any) was eclipsed by his daredevil act of racing a motorcar against a Blériot, a Demoiselle, or one of the Curtiss-type biplanes. Collectively these feckless fellows staged a road show that outran the comedy hit of New York's theatrical season: *The Aviator,* in which a real Blériot nightly took off into the wings on wires. "What kind of oil do you use?" was one of the lines—to which the rueful answer was, "Omega" (a widely advertised liniment).

It would be difficult to conjure up greater trials to the spirit than those that beset the band of airborne gypsies. Often the weather was atrocious; it snowed and hailed in Oklahoma, it was bitterly cold in Texas, and low-lying clouds or fog seemed to be everywhere. Attendance rose or fell with the state of the atmosphere. If it was too breezy for flying, the customers demanded their money back—backing up threats with revolver shots if need be, or (as happened in Houston) burning down the grandstand. County fairgrounds were mostly inadequate, making takeoffs and landings dangerous as well as difficult. Motors were always giving trouble, while crackups were so frequent that at one time the troupe was left with a single serviceable machine. Damaged planes were patched up and put into the air in highly dubious condition. Hotel accommodations varied with the state of the company finances—which in turn depended on the number of paid admissions. But the culminating misfortune, which almost ended the tour, was the fatal accident to Moisant at New Orleans on December 31, 1910.

With his usual flair for the dramatic, Moisant chose the last day of the year—in other words, the last possible moment—to try for the Michelin nonstop distance prize. He would have to beat a French mark of 362.7 miles in 7 hours 48 minutes 31 seconds, set by Maurice Tabuteau only the day before with a Maurice Farman biplane. But Moisant was counting on the luck that had ridden with him so far—as well as on the one Blériot in the stable considered in good enough condition to make the attempt. Flying a short hop from the nearby racetrack, where the circus had pitched its tent, to territory more favorable for the long grind, he was coming down with the wind when a gust upended the tail of his machine. Moisant was pitched forward and out from a height of fifteen feet, breaking his neck. (By coincidence, at approximately the same hour, Arch Hoxsey—twin star with Ralph Johnstone in the altitude events at the Belmont meet—was killed at Los Angeles when he lost control of his Wright and turned over during a "spiral glide" from a great height.) Actually, the winner of the 1910 Michelin trophy was in doubt until late that December 31; for the Alsatian flyer Pierre Marie Bournique, setting record after record for speed with his R.E.P. at Buc, threatened to beat Tabuteau. Bournique covered 330 miles in 6 hours 30 minutes before having to give up.

True to tradition, the show went on after the death of Moisant. John B.'s older brother Alfred took charge; and in the pleasant weather of Mexico and Cuba proficiency rapidly increased, exhibitions were more successful, and gate receipts prospered. When the tour was over and the troupers disbanded early in 1911, Alfred Moisant returned to New York and opened an aviation school at Hempstead Plains, near Garden City, Long Island, where a vast acreage was admirably adaptable to practice flying. Alfred had the assistance of Harold Kantner, an early exhibition flyer, as well as of George H. Arnold, Mortimer F. Bates, J. Hector Worden, and Chief Pilot S. S. Jerwan—"all licensed aviators," as the prospectus put it. A sister, Matilde Moisant, lent glamor to the school by becoming an expert aviatrix, winning a respected place for herself among her male colleagues.

Alfred Moisant summed up the current situation of aviation in the United States accurately enough in an illustrated brochure:

Never has a vehicle so captured the popular imagination and gained so much public support as has the aeroplane. But the United States is today far behind every other modern nation in the development of the flying machine. This should not be so, for we have in this country the finest material for great pilots that can be found anywhere in the world. We have the factories wherein to build aeroplanes and flight motors. As a people, we have native courage, ability, common sense, everything aeronautical. In Europe it is a common thing to fly; here it is a rarity. . . . It has been in the hands of the competent aviators developed by contest flying in Europe, particularly France, since January 1, 1910, that the aeroplane has, to date, made its greatest, if not all its real progress. Two or three Americans there were among these foreign-taught airmen who did what they could to raise aviation in the United States from its condition of stagnation or worse to the place it occupied abroad. But it was the invasion of the United

States by European and European-taught monoplanists last autumn that finally started the pendulum of American aviation to swinging in the right direction.

Moisant proposed to remedy the unhealthy state of affairs at home by teaching pupils, for $750 apiece (plus breakage deposit), "the details of aeroplane construction, repair, adjustment and flight" in five weeks or perhaps a little longer. But he was mistaken as to the direction in which the pendulum was swinging. The first of the great European races took place in May, turning all eyes toward Europe; and the events that followed showed plainly that the United States was losing ground.

When the flying season opened in the spring of 1911, the world was still hesitant in making a choice between biplanes and monoplanes. The biplane, with its greater wing surface area, could presumably carry heavier loads than the monoplane. With the upper wing firmly connected to the lower by means of struts and wires, the biplane also was structurally more rigid, as well as more stable in flight. The problem presented by the monoplane was largely one of structural strength; additionally, while the design was cleaner and hence allowed greater speed, a wing long enough to provide adequate lift required strong external bracings, which interfered with the airflow and created drag. Moreover the single surface, when properly reinforced with the materials then available, usually meant a heavy weight increment.

While America concentrated on the biplane, European constructors devoted time and thought to the challenge of the faster monoplane. Two new machines of particular import made their appearance in France; both were to develop into prize-winning craft. A third model, destined to be supreme in speed contests during the last two years before World War I, was beginning to emerge from the experimental stage.

Out of a near-fatal accident to Léon Morane, the Blériot pilot, and his brother Robert grew the design for an improved monoplane that combined the best qualities of the Blériot construction with greater strength and more speed. While trying for the Michelin prize on October 10, 1910, both brothers broke their legs and suffered other serious injuries when a wing of their two-seater collapsed and they were hurled to the ground with terrifying impact. In the hospital while they convalesced, the Moranes drew up plans for what they conceived to be a better machine, and then employed the talented engineer Raymond Saulnier to build it. The Morane-Saulnier was tested by another well-known Blériot flyer, Emile Aubrun—who had taken second place in the Circuit de l'Est, had barnstormed in South America, and was the holder of the first license granted by Argentina. The new monoplane soon became one of the Blériot's greatest competitors.

Another newcomer to the monoplane fold was the Nieuport, a machine of carefully streamlined form, with a rectilinear fuselage of steel tubing

The original Nieuport monoplane in flight at Issy-les-Moulineaux. To streamline his later models, Nieuport rounded the nose and flattened the wing curve.
(*National Air and Space Museum*)

covered with fabric and a rounded nose that effectively reduced drag. Edouard de Niéport—a name which became Nieuport—was born in Algeria in 1875, the son of a colonel in the French army. Against the wishes of his father, Edouard gave up studies at the Ecole Supérieure d'Electricité to become a racing cyclist; subsequently he turned his efforts to the manufacture of magnetos, spark plugs, and electrical appliances. In 1908 he took up aviation at Issy-les-Moulineaux, where he contrived to get into the air with his first model, powered by a small Darracq motor. Studying the situation from an engineering standpoint, he decided in 1910 that the monoplane with covered fuselage was the plane of the future. First equipped with a 28-hp, twin-cylinder engine of his own construction, Nieuport's machine incorporated the fortuitous combination of a wing curve flatter than any yet seen and a carefully contoured body—which soon enabled him to break records at Mourmelon. He had discovered the principle that the less the camber (curve) of the wing, the faster the speed. With the later substitution of the more powerful 50-hp Gnôme, his monoplane became an emblem of swift flight. Nieuport's influence on research and design was far reaching. There is no telling where his genius in aerodynamics might have led if his life had not been tragically cut short at the moment he achieved success.

Equally promising was the Antoinette-type monoplane of Armand Deperdussin, an enterprising and generous silk merchant who employed as his designer one of the truly great engineers of the aeronautical world, Louis Béchereau. The prototype of the Deperdussin was a four-bladed *canard* monoplane designed in 1909 for Christmas exhibition in a Paris department store. The first flying model, built in 1910, performed well from the start. Powered by a 40-hp, four-cylinder Clerget engine, it had a

Silk merchant Armand Deperdussin sponsored one of the first planes to be steered by "wheel" rather than by "stick." He also added wheel skids in front of the long, narrow fuselage. A 1911 model is shown here. (National Air and Space Museum)

long fuselage with a very small cross section, two wheels with skids, and a sturdy tail skid. It was one of the first machines to employ the "wheel" control, as distinct from the "stick" control—an innovation sometimes referred to as the Dep control. While some early pilots complained that it had an irrepressible tendency to steer to the left—to "chase its tail"—Béchereau's advanced construction was on the threshold of worldwide fame.

American monoplanes were far behind European designs of the period. Dr. Henry W. Walden in 1909 had built and flown a pusher type, with tricycle landing gear, that resembled a Curtiss biplane minus the upper wing. Eventually its inventor found a place in the National Aviation Hall of Fame—"for conceiving, building and demonstrating the first successful monoplane in the United States." A tractor monoplane built by Walter L. Fairchild at Mineola in 1911 pioneered the use of steel-tubing construction in this country. Other than these, however, American contributions at the time to aeronautical development consisted not in monoplane design but in the Wright and Curtiss biplanes and their sundry variations and imitations.

In March of 1911 an ambitious attempt was made by Claude Grahame-White and Harry Harper, coauthors of *The Aeroplane, Past, Present and Future*, to compile a list of those men and women known to have flown successfully up till then. They set down a total of 729 names. Of the six nationalities providing the majority of aviators, the Americans were last —with only 31 pilots. The score for France was 387; for England, 128; for Germany, 46; for Italy, 38; and for Russia, 37. Three years earlier, at the beginning of 1908, the Wrights had been almost alone in the field.

A breakdown of the 729 aviators showed that 361 used biplanes, 302

Students enrolling in the Farman flying school at Etampes, France, were "graduated" after three solo flights. Photo on this page shows the main entrance to the school; opposite, an outdoor lesson is in progress.

preferred monoplanes, and 66 flew aeroplanes of "original design." Further analysis verified the overwhelming predominance of Henry Farman's biplane and the even more popular Blériot monoplane.

Principal Biplanes		*Principal Monoplanes*	
Farman, Henry (French)	135	Blériot (French)	158
Voisin (French)	59	Antoinette (French)	47
Wright (American)	39	Hanriot (French)	21
Sommer (French)	36	Demoiselle (French)	10
Bristol (British)	16	Tellier (French)	10
Farman, Maurice (French)	14	Koechlin (French)	8
Aviatik (German)	14	Valkyrie (British)	7
Curtiss (American)	13	Gyp (French)	5
Other	35	Nieuport (French)	5
Total	361	Etrich (German)	5
		Morane (French)	4
		R.E.P. (French)	4
		Other	18
		Total	302

While it would be disingenuous to pretend that these figures are exhaustive (there was no way of telling, for instance, how many amateurs were flying without license), the compilation is nonetheless illustrative of the fantastic burgeoning of aviation in little more than seven years and of the extent to which Americans had been left behind.

The Henry Farman biplane that proved such an emphatic favorite with aspiring pilots in 1911 sold for 28,000 francs, with a choice of Gnôme, Renault, or E.N.V. engine—or for 13,000 francs without motor, propeller, and gasoline tank. A good example of the nearly universal popularity of the Henry Farman model may be found in the story of the first plane to fly in Japan. In May 1910, Yoshitoshi Tokugawa, a twenty-seven-year-old first lieutenant in the Japanese engineer corps, was among a group of likely candidates selected by various governments to study aviation in France. Together with students from Germany, Poland, and Russia, he entered the Farman school at Etampes, where he obtained his license on November 8. Instruction then consisted of ten five-minute training flights at an altitude of about thirty meters. The pupil was wedged in at the instructor's back and reached over his shoulder to hold the stick. With a rotary motor the pilot had to allow for the resulting torque, or tendency to veer from a straight course. Consequently, left-hand turns were easier than right; and these were the only kind allowed. The process was somewhat comparable to learning how to ride a bicycle. After just three solo flights the student was "graduated."

Tokugawa brought back to Japan a Henry Farman two-seater with 50-hp Gnôme motor, which he had purchased for the sum of 18,832 yen, and in addition a Blériot costing 15,602 yen. In 1950 he recalled in a published article the perplexities of assembling the Farman at the Nakano Balloon Corps in Yokohama. Once the wings were in place, it proved impossible to disassemble them; and a squad of fifty men was needed to transport the machine to the Yoyogi Maneuver Ground in Tokyo for a test flight. With Tokugawa—by now a captain and an "experienced" aviator, after Etampes—at the controls, the plane became airborne on the

Aerial view of the hangars at the Farman flying school.

morning of December 19, 1910, flying 984 feet at an altitude of 230 feet. On April 23, 1911, Captain Tokugawa set a Japanese record with the Blériot, flying 48 miles in 1 hour 9 minutes 30 seconds.

These pioneer flights antedated by many months the appearances of Thomas Baldwin and other American flyers who began touring Japan for exhibition purposes in the latter part of 1911. The Farman survived until March 28, 1913, when it came apart in mid-air, killing the occupants. After World War II it arrived in the United States in fragments as part of a shipment of intelligence material, and was subsequently restored by Air Force Museum personnel at Dayton. In an impressive Armed Forces Day ceremony in 1960, fifty years after its first flight, the biplane was formally returned by the United States to the Japanese government as a gesture of good will. Tokugawa, by then a general seventy-seven years of age, sat again at the controls and told his audience that he felt "just like a father . . . moved to tears to see his brilliant son come back to Japan after completion of his studies in the United States."

To another historic Henry Farman belongs the distinction of being the first aeroplane to fly in Asia. Piloted by the Belgian Charles Van den Born, it achieved that honor at Saigon on December 15, 1910. A Henry Farman plane in the hands of Germano Ruggerone, an Italian flyer who gave exhibitions in South America, was also, in 1910, the first flying machine ever seen in Brazil. Still another Henry Farman was taken to India by the French pilot Tullerot in 1911; and its close cousin, a Sommer, was the first to invade the air of China, in 1910—guided by another Frenchman, R. Vallon.

A Farman derivative, the Bristol Boxkite, also was successfully exported. Manufactured by the British and Colonial Aeroplane Company, this slow-moving but simply operated biplane saw service in such distant countries as India and Australia. It became a standard training machine for the British, and later the Russian, armies. The Boxkite was developed by Sir George White, millionaire owner of the Bristol Tramways Company—who, it was said, located his flying field at nearby Filton in order to stimulate passenger traffic on that suburban transportation extension. Before the familiar model passed out of existence, 130 Boxkites were built.

Among the monoplanes, the marked success of the Blériot was due not only to its popularity in France but to the fact that foreigners were disposed to fly it as well; moreover, it was beginning to find acceptance by military pilots. Insofar as their own manufacturers were unable to meet production requirements, sales were made to the British, Russian, Italian, and Austrian governments in addition to the French. Such remote private flying clubs as the Ligue Nationale Aérienne de Saigon and the Aéro Club de Sebastopol likewise acquired Blériots. Competing sportsmen and prize seekers who were in the market for a monoplane found the 50-hp Blériot-Gnôme combination irresistible because of the machine's speed and fast rate of climb. The one-seater model sold in Paris for 24,000 francs ($4,800), the two-seater for 28,000 francs—compared with their Anzani-powered counterparts built by Moisant in the United States, where the 30-hp one-seater went for $3,100 FOB; the 50-hp one-seater for $6,200; and the two-seater model for $6,500.

Admittedly it was expensive to learn to fly. Clifford Harmon, conversing with friends at Mineola, pointed out: "Your aeroplane costs more than five thousand dollars. You must pay a fee of five hundred dollars for tuition, and you must deposit, too, about five hundred dollars more for damages. All the damage you do to the machine you learn on must be paid for, and your deposit may easily be eaten up." Harmon told of an English flyer whom he had met at Nice. "I learned to fly in a week," Harmon said to the Englishman. "How long did it take you?"

"Oh, nine or ten. . . ."

"What!" Harmon interrupted. "Not nine or ten weeks?"

"No, aeroplanes," the Englishman replied.

The Blériot was the chosen instrument of many other pilots, who ranged far and wide to show the people of other lands that the aerial age had arrived. Blériot himself was the first to ascend at Budapest, on October 17, 1909, and at Bucharest, on October 30—a few weeks before his near-fatal crash on a rooftop in Constantinople.

Yves Guyot, a Frenchman who attempted to demonstrate his Blériot to the Russians, encountered official difficulties at St. Petersburg. A special cable to the *Washington Post* dated December 11, 1909, reported that all attempts at aviation there or in other fortified areas were expressly forbidden. A well-known Russian publisher who sought to raise subscriptions to a "national airship fund" was warned that his proceedings rendered him liable to suspicion, "as aerial craft might be used by revolutionaries and other evil-disposed persons. Any airship or aeroplane observed within the prohibited area" would be fired upon without notice. Guyot, however,

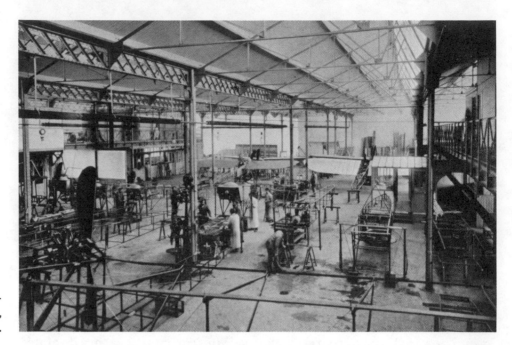

Blériot factory work-room at Levallois, France, 1911.

was ultimately allowed to take off; he demonstrated at St. Petersburg and Moscow in November 1910.

Scandinavia made the acquaintance of the Blériot through the trail-blazing flights of the Swedish pioneer Baron Carl Cederstrom. In stature and spirit a typical Norseman, Baron Cederstrom found an outlet for his love of sports by obtaining a license in France in 1910, then purchasing a Blériot and giving demonstrations in Sweden, Norway, and Denmark. The "Flying Baron" and his monoplane, the Bilbol, became a familiar sight throughout the Nordic countries. A friend recounted one instance of the Swede's daring at the Amager flying field, near Copenhagen:

Coming down at the far side of the aerodrome, Cederstrom was too impatient to wait until mechanics could get across to start his engine, so he jumped out of the machine, leaving the switch on, and swung his propeller. Diving under the wing as the machine gathered speed, and catching hold of the fuselage . . . which was by then already moving along at a good pace, he swung himself aboard and rose gracefully into the air. This performance was repeated not once, but several times.

This performance, it may be added, was attempted by others—sometimes as a stunt, sometimes of necessity—but not always with such success. It sometimes resulted in a riderless plane running amok, until brought to earth with greater or less damage, as the case might be. Cederstrom lived to pursue his antics, and eventually to establish a factory and flying school —until July 1918, when he was lost in the Baltic with his chief instructor while delivering a hydroaeroplane to Finland. His body was recovered and taken to Stockholm for burial.

Spain and Portugal were both invaded by Frenchmen flying Blériots.

Hubert Leblon made a series of flights at San Sebastián, where, on April 2, 1910, he crashed on rocks hidden beneath the sea and drowned in the wreckage. Lucien Mamet, an early associate of Blériot, flew in Portugal in April after giving the inhabitants of Barcelona, Spain, their first glimpse of a plane in flight on February 10. Meanwhile, the Belgian Jan Olieslaegers, another rising star, demonstrated the Blériot at Seville and at the North African port of Oran, Algeria. Cattaneo, the Italian pilot, was a busy Blériot pioneer in South America; on December 17, 1910, he won a prize equivalent to $20,000 for flying 70 miles across the river La Plata from the Argentine to the Uruguayan side and back. The French-man Jean Raoult carried the Blériot standard to Madagascar, where he made an abortive attempt to start a mail service. A Russian military pilot, Lieutenant A. de Kouminski, was the early bird, with another Blériot, in Persia, Siam, and Cambodia.

If it wasn't a Henry Farman or a Blériot, it was apt to be a Voisin that first found a path through foreign skies. Within a few months after the Rheims meet, Voisin Frères had opened a sales office in Paris, where they advertised their product as "the fastest and most stable of all air ma-chines. . . . The buyer pays only after taking one flight in the aeroplane he intends to buy." Henri Brégi, chief pilot of the Voisin school, was sent to South America to open an export unit—and soon the big biplane was well known to wealthy sportsmen. One of these, Alberto Braniff, was the first to make a flight in Mexico. (Alberto was a remote cousin of Thomas E. Braniff, who in 1928 founded Braniff Airways, to operate between Oklahoma City and Tulsa. Both were descendents of an expatriate Irish family, one branch of which had settled in Mexico and another in the Uni-ted States.)

With the exceptions of Baldwin and Bud Mars, few Americans were found among those who advertised the marvels of flight abroad. Mars was the first foreigner to fly in Japan; he performed in his Curtiss before a huge crowd at Osaka and gave the young Crown Prince Hirohito his first ride in an aeroplane in 1911. He also toured the Philippines and Ko-rea. On July 18, 1909, the Wright biplane of Eugène Lefebvre became the first to fly in Holland. Another Wright, manufactured under license in Germany, became the earliest flying machine to make public flights in Switzerland; on March 13, 1910, the German flyer Paul Engelhard took off from the frozen surface of the lake at St.-Moritz. The wisdom of this experiment had seemed open to question at first, because of the thin atmo-sphere at the 6000-foot altitude. However, it proved a success—with an added advantage in that a ring of snow-clad mountains afforded protec-tion from the wind.

A 1910 compilation showed that about seventy types of engine were being constructed at the time in the United States and Europe. The Wright water-cooled, four-cylinder vertical engine, producing up to 40 hp; the Curtiss 60-hp, six-cylinder engine; and the pioneer V-8 of Curtiss (with which he won the Gordon Bennett race) were among the most widely used in the United States. Almost all of these motors were derived from the stationary automobile engine. A miscellany of similar makes (such as the water-cooled, V-type Hall-Scott; the Sturtevant; and the Thomas Morse) were fitted into experimental planes here and there. Most engines had four-cycle operation—although the two-cycle, six-cylinder vertical Roberts engine, producing up to 100 hp, was soon to be built in considerable numbers. The Kirkham and Christofferson vertical engines also were to become well known in the years ahead.

In Europe the Gnôme rotary engine was supreme. Specially designed and built for aviation purposes by the Séguin brothers, Laurent and Louis, this light seven-cylinder motor with air-cooled flanges—the 50-hp model weighed only 176 pounds—was installed so as to revolve with the propeller around a fixed crankshaft. The use of an odd number of cylinders was based on Esnault-Pelterie's idea for combining light weight with power. The first model had five cylinders; soon seven-, nine-, and fourteen-cylinder models were being built. The 50-hp Gnôme delivered from 200 to 1300 rpm (according to one user, 900 was considered par) and was relatively easy to take down and put together again—a great help in the days when valves and spark plugs had to be cleaned frequently. The crankcase of the Gnôme was like a steel hoop, while the steel crankshaft was hollow; gasoline and oil passed through it to the cylinders by centrifugal force. Overheating, on the one hand, and ice formation in the carburetor, on the other, were nearly unknown; for the rotary engine kept itself cool by its spinning motion, while carburetion took place mostly inside the warm crankcase.

As in other aeroplane engines, castor oil was used for lubrication because it was not so readily cut by gasoline. But the Gnôme was a notorious consumer of both; and gradually it came to be supplanted by the more economical Le Rhône. An informal comparison made on the eve of World War I indicated that the Gnôme used roughly ten gallons of gasoline and two of oil to the Le Rhône's six gallons of gasoline and only one of oil. Other rotaries included the Clerget in France and the BMW (Bayerische Motoren Werke) in Germany. For a time the three-cylinder radial Anzani was a rival of the Gnôme; but like the R.E.P., its low horsepower proved inadequate for the growing power demands of more modern aeroplanes.

The primacy of the V-type, eight-cylinder Antoinette, the original power plant of European aviation, was later challenged by two stationary engines. Designed on the same general lines was the heavy (396 pounds) Renault V-8 of 50 to 60 hp, whose cylinders, instead of being cooled by water, had a simple arrangement of flanges exposed to a strong draft created by a fan. It seemed a natural adjunct to the Maurice Farman biplane, and many pilots thought it more dependable than the Gnôme. The even heavier water-cooled British E.N.V. was used principally in the Voisin, by such pilots as Rougier, de Caters, Métrot, Duray, and Efimoff. At the 1910 Heliopolis meet, E.N.V.-powered planes accounted for a total of

145,000 francs in prize money, out of a possible 187,000 francs. The name E.N.V. had a curious origin. Although British crafted and financed, this engine was made at a plant in the suburbs of Paris, the owners having realized that the best market lay in France. Known to the French as *le moteur en V* ("motor in V form"), the appellation metamorphosed into the initials "E.N.V." in English; this expression was adopted as the company name thenceforth. In Germany and Austria, the vertical-type Mercedes-Benz was a heavy favorite, while in Italy one of the most popular motors was the similar Isotta-Fraschini.

As the spring of 1911 drew near, a plethora of fresh prizes lured Europe's aerial buccaneers into new efforts. To start the season, Robert Grandseigne made the first night flight over Paris, on February 11, 1911, in a Caudron biplane outfitted with electric light bulbs. Then Eugène Renaux, a former racing-car driver, won 100,000 francs on March 7. In a Maurice Farman biplane and carrying a passenger, he made an extraordinary flight from Paris to the Puy-de-Dôme, a mountain in southwestern France, on whose summit—a tiny plateau only 150 yards square—he executed a skillful landing. The 360-km distance was covered in 5 hours 10 minutes 46 seconds—50 minutes under the permitted time. Renaux was to win another 30,000 francs on October 1, carrying a passenger in the Quentin-Bauchart contest, for the greatest distance flown in a given period of time. There were few equivalent statistics, at the time, in the United States. Records were not being made by Americans; whether European goals were too high for the capabilities of existing machines or the experience of their pilots, however, remained a question.

Bold new projects were announced—the greatest races yet: from Paris to the distant capitals of Spain and Italy; around the compass of central Europe; around the compass of Great Britain; around the compass of Belgium; around the compass of Germany. Confidence had so increased that instead of merely circling an aerodrome or hopping from town to town, flyers essayed trips across the whole territory of a nation. Before the aeroplane, international boundaries vanished and new prospects for travel opened up. With *Le Matin* amply compensated in popular interest —and circulation—by its Circuit de l'Est, *Le Petit Parisien* followed suit, putting up 150,000 francs for a race from Paris to Madrid; the *Petit Journal* offered prizes totaling 500,000 francs for a race from Paris to Rome; the *Journal* nearly as much for a "Circuit of Europe"; and the *Daily Mail* almost again as much for a "Circuit of Britain." Including prizes awarded in the United States and elsewhere, it was estimated that more than $1,000,000 was earned in the 1911 flying season.

Long before the gray, gusty dawn of Sunday, May 21, a crowd estimated at three hundred thousand persons began to converge on Issy-les-Moulineaux to watch the start of the Paris–Madrid contest. Unending lines of automobiles and horse cabs snaked their way through the ranks of pedestrians. Intent on witnessing the takeoffs of twenty-nine entrants, the spectators half expected to view spectacular crashes.

They saw shockingly more than they had anticipated. While getting off the ground in a monoplane of his own construction, Emile Louis Train lost control when his motor faltered. With the propeller still turning, he plunged into a group of high officials, killing Minister of War Berteaux and inflicting injuries on many others, including Prime Minister Monis. The catastrophe could have been avoided. Impatient spectators had spilled over from the edges of the field, and the official party, under escort of a detachment of Horse Guards, found itself in the path of planes taking off. The unfortunate Train managed to clear the horsemen but fell into the middle of the delegation.

The Paris–Madrid event brought to the fore one of the all-time champions of flight—Jules Védrines. A man of tempestuous temperament and great physical endurance who formerly had served as mechanic and chauffeur to the actor-aviator Robert Loraine, Védrines first attracted notice as the pilot of a Goupy biplane—a small but excellently made tractor with open fuselage and staggered wings. He was the short, swarthy son of a Paris working-class family, and an avowed socialist. A chronically aggressive flyer, Védrines obtained his license at the Blériot school in Pau on December 7, 1910. Like many less well-to-do aviators of the day, Védrines was employed by the manufacturer of the plane he flew—the new Morane-Saulnier monoplane.

In the melee that followed the Train disaster, Védrines aborted his takeoff and damaged the machine. The race was postponed until the following day, and in the interval Jules borrowed from a friend another version of the same basic monoplane—a Morane-Borel. Shaken by the bloody accident or beset by mechanical difficulties, all but six of the entrants had canceled out by the time the race was started again on May 22. Only four managed a clean getaway: Védrines; Roland Garros, then virtually unknown in France; Eugène Gibert, a former university athlete and garage mechanic who had taken his pilot's license at the age of twenty-one; and Jean Conneau, a naval ensign flying under the mellifluous pseudonym "André Beaumont." Garros, Gibert, and Conneau flew Blériots; all four participants used Gnôme motors of 50 hp. Conneau, who had resolved to develop his knowledge of flying for the benefit of the French navy, was eliminated when he cracked up on the easy first leg from Paris to Angoulême, leaving Garros in the lead.

The second stretch was far more difficult—from Angoulême over the forbidding Pyrenees to the Spanish seaside resort of San Sebastián. During this lap Védrines forged ahead. Both Védrines and Gibert claimed that they were viciously attacked by eagles while negotiating the high, dangerous pass of Somosierra—the former escaping through evasive tactics, the latter by firing his revolver. Whatever the facts of the encounter, it furnished a fabulous story for the *Petit Parisien*.

On the last leg, from San Sebastián to Madrid, Gibert and Garros were eliminated by motor trouble, and Védrines finished alone—having covered a total distance of 842 miles in 14 hours 55 minutes 18 seconds of flying time. Throughout the race Védrines had driven his machine as if possessed. Upon his landing at Garafa, outside Madrid, where a crowd of a hundred thousand wildly adulatory Spaniards had assembled, the welcoming committee had good reason to believe that the devil had taken charge. Instead of acknowledging the cheers, so the story goes, Védrines sat scowling and cursing at the imaginary injustices visited upon him by Providence during the long flight. Fearing that his mind had become unhinged, the committee sent for a doctor—which only added fuel to his fury. It was some time before the hero of the occasion calmed down sufficiently to receive the congratulations of King Alfonso, who had been an ardent advocate of the aeroplane since meeting Wilbur Wright at Pau.

Despite his frequent rages, Védrines became a public idol. He was to perform many daring feats in the course of his career—including a flight from Paris to Cairo in 1913. After World War I, he took credit for one of the most spectacular stunts of the day. The Galéries Lafayette, a Paris department store, had offered a prize of 25,000 francs for the first flyer to land his plane on its rectangular roof, which was only 20 meters long by 12 meters wide. Védrines accomplished the seemingly impossible—not with a helicopter but with a small, slow Caudron biplane, which he maneuvered to a three-point landing.

Védrines's flying style had a strange effect on his rivals. "He was always at my heels, pursuing and menacing," Conneau was to write later. "I heard him in the air as I drove over forests, mountains, and lakes; I heard him at the landing-places where he had descended—usually soon after I had landed. After a while I heard and saw him in my sleep. Somehow he obsessed me, creating a sort of fear in me, such as I had never experienced before."

"André Beaumont" had himself something of a Mephistophelian appearance, with his sharp black eyes, finely trimmed black beard, and waxed moustache—especially in his flying suit with black-visored woolen cap. And in the next race his steady, accident-free flying was almost diabolical in its cleverness.

The Paris–Rome contest, starting May 28, was originally Paris–Rome–Turin; but the Turin leg was amputated to become a separate member, after a week's interval, as Rome–Turin. The *Petit Journal* had provided prizes for the order of arrival at Nice and at Rome, as well as at Turin, the winner assured of a grand prize of 300,000 francs. Machines or parts of machines could be substituted at will, provided notification was given to one of the control points. This time the crowds were forced far out of town to the start at Buc, beyond Versailles. Nevertheless roads were again choked with vehicles; expectant hordes slept, sang, or picnicked through the night, while detachments of gendarmes and cavalry took up positions to prevent a repetition of the tragedy at Issy. At 5:30 A.M. biplanes and monoplanes were herded side by side, pilots at the

Flying under the pseudonym "André Beaumont," Ensign Jean Conneau of the French navy became an instantaneous hero when he won the Paris–Rome race of 1911—the first person to fly between the two capitals. In the same year he chalked up victories in the Circuit of Europe and the Circuit of Britain.

ready, motors tuning up; at 6:00 a smoke bomb announced that the race was on. Of the twenty-one aviators who had entered, twelve left the ground.

In the next few days machines were pushed to the limit—for even minutes counted in the scoring. One by one the pilots were compelled to give up: Albert Kijmmerling, Léon Bathiat, and Michel Molla in Sommer machines; Weymann in a Nieuport; Bielovucic in a Voisin; Gaget in a Morane; Romulo Manissero in a Blériot. At length Garros and Conneau—both with Blériots—were left fighting it out neck and neck. It was the careful and consistent Conneau, steering a sure course, who reached Rome first. He arrived at the Parioli racetrack on May 31, in 82 hours 5 minutes elapsed time—to the surprise of the promoters, who had believed the race would last at least a week.

Conneau was met by a surging mob that knocked down women in its eagerness to hoist shoulder high the first person to fly to the Eternal City from the capital of France. The scenes that followed rivaled Blériot's reception in London. There was an audience with King Victor Emmanuel; Pope Pius X raised his hand in benediction of the flight from a window in the Vatican; and emotion ran high among the impressionable residents of Rome.

Garros arrived the following day, after smashing two machines en route, having completed the trip in 106 hours 15 minutes. Only two others stayed the course: André Frey, in 156 hours 52 minutes, on one of the new Moranes; and René Vidart, a cheerful youngster who was flying the slim, swift, fastidiously designed Deperdussin in its first public appearance. But Vidart was far behind, with 195 hours 8 minutes. Attempting to complete the extra course to Turin, Frey crashed on June 13 near Ron-cigillone, France, breaking both arms and legs; this unofficially brought an end to the contest.

The public thirst for country-to-country races seemed unquenchable. An intoxicating draft was the 1600-km Circuit of Europe, under joint auspices of the Paris *Journal* and the London *Standard*. Such affairs were beginning to take the aspect of Roman holidays. Always there was the chance of an accident; and the curious came with just that in mind. When the European circuit began on June 18—again drawing a Sunday crush—such morbid thoughts were again fulfilled.

The military parade ground at Vincennes was all too easily reached by an estimated seven hundred thousand Parisians, who poured out for the 6 A.M. start. Nothing in the history of aviation had ever drawn such an attendance: a solid phalanx of automobiles could make only the barest headway through the army of men, women, and children who marched through the streets in the dawn's first light. Even the pilots had difficulty getting to their hangars.

No fewer than fifty-two machines were entered for the exacting course of 1600 km—from Paris to Liège to Spa to Utrecht to Brussels to Roubaix to Calais to London, and back to Calais and Paris. Practically every type of aeroplane in Europe was represented—although, as in the other great races, not a single American machine took part.

Lieutenant F. Princeteau of the French army, a recent graduate of the Blériot school at Pau, was the first fatality. Attempting to take off from Issy to fly to the starting line at Vincennes, his monoplane capsized and burst into flames. Unable to extricate himself, Princeteau was burned to death before the eyes of his helpless friends—the first to die in a crash followed by fire. Shortly after the start of the race, T. Lemartin, an instructor at the same school and the first to fly the Blériot four-seater, crashed from a height of 200 feet. Two rows of flying machines had been arbitrarily drawn up to stage a dramatic departure, and some who witnessed the accident believed that Lemartin's death might have been avoided had he not taken off in a crosswind. The third to fall was Landron, pilot of the successful monoplane built by Alfred de Pischoff. Landron crashed at Château Thierry, was pinned down by the wreckage, and—like Princeteau—perished in the flames from a ruptured gasoline tank.

Only nine planes managed to complete the circuit back to Paris. Of these, an R.E.P. was the only one to do so without major repairs—whereas some flyers changed planes four or five times, and motors as many as seven times. This machine was also the only entrant whose motor had been conceived and built by the plane's manufacturer. The R.E.P. did not finish in the money, however. Again the winner was Conneau, who led the field throughout, with 58 hours 38 minutes of flying time. He was followed by Garros, the "eternal second," with 62 hours 17 minutes. Once

The first stage of the Circuit of Europe as portrayed by a French newspaper. An estimated 700,000 Parisians gathered at Vincennes for the 6:00 A.M. start of the race.

more the Blériot had scored a double victory. Third place went to Vidart and the Deperdussin, with 73 hours 32 minutes; Kimmerling, in a Sommer monoplane, was fourth.

On July 1, the third Gordon Bennett contest, with a longer course—150 km (94 miles)—than those previous, was held at Eastchurch, on the British Isle of Sheppey. America was represented by "Charley" Weymann, an aviator who had never flown in his own country but whose name was well known abroad. Weymann did not present the usual picture of an aviator: with his clean-shaven, rounded features, wide smile, rimless pince-nez glasses, and loose-fitting raincoat over white turtleneck sweater, he had the casual air of the gentleman sportsman. Already experienced in cross-country flying, he had recently gone in for altitude; and now, with a machine that promised much—the Nieuport—he was ready to try for speed.

To meet the competition of this new monoplane with its powerful 100-hp Gnôme, Alfred Leblanc, mainstay and mentor of the Blériot team (which used the same type of motor), reduced the wing surface of his machine by clipping the ends of its wings—an expedient never tried before. Edouard Nieuport himself, in a duplicate of Weymann's machine but with a 70-hp Nieuport motor; L. Chevalier, with a smaller, two-cylinder Nieuport; and Alec Ogilvie, flying for Britain in a Howard-Wright biplane, were the other principal contenders.

The Gordon Bennett course was flown by Weymann in 1 hour 11 minutes 36⅕ seconds, at the record rate of 78.77 m/hr. Close behind were Leblanc and Nieuport, with 1 hour 13 minutes 40⅕ seconds and 1 hour 14 minutes 40⅕ seconds, respectively. It was proof, if further proof were needed, that the monoplane could be expected to excel in speed: Ogilvie's biplane was a slow fourth, in 1 hour 49 minutes 10⅖ seconds.

More particularly, the results showed how closely the Blériot was being pressed. Gustav Hamel of the British team had also shortened his wings in the hope that his Blériot would outpace the Nieuports—but the results were calamitous. Rounding a pylon, the youthful, curly-haired Hamel sideslipped and crashed in a cloud of dust, demolishing his machine—fortunately without injury to himself.

Although the trophy had been temporarily retrieved for the United States, the 1911 Gordon Bennett tourney was in fact one more victory for the manufacturers of French planes and motors. Nonetheless it gave America another chance to play host and defend its possession of a leg on the cup.

Hardly had the victory of his machine been celebrated in print and picture than Edouard Nieuport was killed. During the French military trials at Charny on September 16, he glided down in a series of sharp turns with the motor cut off; caught by the wind at the moment of alighting, he switched on the power. But his Gnôme—flooded with gasoline—failed to respond. In a crash landing, the pilot was thrown heavily forward, suffering fatal injuries.

Edouard's brother Charles, with whom he worked in close partnership in the tradition of other such teams, was also to die in a crash, with his mechanic, at Etampes on January 24, 1913—due to a malfunctioning of the plane's wing-warping device. But the name of Nieuport carried on, to achieve an enviable reputation as one of the best-known of all fighter aircraft in the early stages of World War I. The only surviving example of the first war model, it is believed, stands in the museum at Linköping, Sweden.

Next on the crowded calendar of 1911 came the Circuit of Britain, for which the *Daily Mail* had offered a first prize of £10,000. A new test was posed for the competing aircraft: instead of freely permitting replacements, as in the case of Paris–Rome and other races, the rules provided that five essential parts of each machine, to be stamped by the Royal Aero Club, could not be removed or changed during the contest. The object of course was to bring about improvement in design—one illustration of the value of such competitions.

On a sizzling-hot Saturday, July 22, seventeen out of twenty-eight entering pilots finished the first brief stage, from Brooklands to Hendon—a variegated procession that included Blériots, Bréguets, Bristols, a Grahame-White Baby, a Cody, a Deperdussin, a Morane-Borel, a Nieuport, a Blackburn, a Birdling, and a Howard-Wright—all cheered by enormous crowds at each end. To avoid a Sunday mob, it was decided to run the second stage—to Harrogate, Newcastle, and Edinburgh—on Monday, July 24; but before dawn that morning a long line of rumbling automobiles, motorcycles, and taxis was already threading a path through those on foot, in a massive assault on the field at Hendon.

The race was a hard-fought, touch-and-go match between Védrines and Conneau. Only two machines disputed Conneau's seemingly invincible Blériot: the Morane-Saulnier of Védrines and a Deperdussin flown by James ("Jimmy") Valentine, the sole Englishman to figure in the running. Valentine had gained his certificate at Brooklands, on January 17, 1911, in a Gnôme-powered biplane built by Robert Macfie; but he was soon attracted to the clean lines and equally clean performance of the French challenger in the monoplane class. It was his misfortune to be put out of the contest by a broken propeller.

Careering from Edinburgh to Manchester and then to Bristol, by way of Glasgow and Stirling, the duel between the two Frenchmen continued in suspense. It was best described by Ralph Simmonds in his book *All About Airships:*

You can picture those two goggled, muffled airmen, crouching over their wheels, blown this way and that by furious mountain gusts in the gorges of Scotland, so that at times their aeroplanes barely moved; tossed like leaves in a gale; lashed by rain and hail that fell from the grey skies; air-sick, numbed and weary, yet holding steadfastly to their work, listening anxiously to the beat of their engines, and operating their controls with deft hands as they drove over hill and dale—southward ho! A more wild and picturesque race than this has surely never been won.

Conneau's naval training and his skill as a navigator—his familiarity with charts, compass, and wind—were invaluable assets. In front of him in the open cockpit of his Blériot, he had one of the first aerial maps on rollers, as well as a clock, compass, and altimeter—the best (even if rudimentary) equipment available at the time. And he made good use of it all. The impetuous Védrines, on the other hand, lost time by losing his way. Again Conneau was the winner; with Védrines at his heels, he showed the British how to fly 1010 miles in 22 hours—at an average speed of about 50 m/hr.

Conneau's capture of the three great prizes of the year was naturally good copy. *Mes Trois Grandes Courses* ("My Three Great Races")—profusely illustrated with photographs of the author, describing each of the races in detail, and summing up the pilot's impressions, was rushed off the presses. As might be expected of a sailor, Conneau stressed navigation above all—particularly the importance of using a compass, an instrument with which few planes were then supplied. The book was the springboard for a series of Paris lectures in the winter of 1912, in which "André Beaumont" proved himself as polished a speaker as he was an aviator.

With the example of France before them, the Germans could no longer afford to concentrate on Zeppelins to the exclusion of contests for flying machines. A "Circuit of Germany" adhered to the prevailing pattern. Fourteen starters left Berlin on June 12, under sponsorship of the *Zeitung am Mittag,* for a month's journey—arranged in thirteen stages—over a total of 1850 km. Wind and weather defeated most of them: only three completed the course back to the German capital by July 10. First prize went to Martin König, in an Albatros biplane; second to Hans Vollmöller, in an Etrich Taube monoplane; and third to Bruno Büchncr, in an Aviatik.

The Belgians, too, caught the contagion. A "Circuit of Belgium" was organized for August 6 to 15—a race of some thousand kilometers in five stages, with ten of the original twenty entrants crossing the starting line. The winner was Jules Tyck in a Blériot; second and third places were taken, respectively, by Count Joseph d'Hespel and Alfred Lanser, both in Deperdussins. Continet, in a Wright-Avia biplane, and Parisot, in a Henry Farman, were the others to finish.

In the United States, where meets were fewer and prizes less attractive, aviation continued to expand during 1911—but at a much slower pace than in Europe. Calbraith Perry Rodgers (a direct descendent of Commodore Matthew Calbraith Perry, who opened Japan to trade, and of Commodore Oliver Hazard Perry, hero of the Battle of Lake Erie) flew 4321 miles from coast to coast—the longest distance yet covered by an aviator. The flight was almost literally crash by crash, with seventy unhappy landings; the only parts of the original machine intact at the finish were one strut and the rudder.

Rodgers was competing for a handsome prize of $50,000 that William Randolph Hearst—never one to miss a chance for the front page—had announced in the *Los Angeles Examiner* by way of a local buildup for a meet of California aviators in October 1910. The money would go to the first flyer crossing the North American continent in 30 days; he could stop as many times as he liked, but the trip had to be completed before October 10, 1911. The one-year period of grace had nearly lapsed before anyone ventured to enter the transcontinental derby.

First off was Robert G. Fowler, whose departure from Golden Gate Park in San Francisco on September 11, 1911, was cheered by a crowd of ten thousand people. But the windy heights of the rugged Sierras proved too much for his Wright biplane—whose four-cylinder, water-cooled engine, he said, "sometimes" delivered 32 hp. He shipped the machine—named the Cole Flyer for his sponsor, the Cole Motor Company of Indianapolis—to Los Angeles for another try.

On September 13, James ("Jimmy") Ward, a football-helmeted eighteen-year-old who had received his pilot's license barely a month before, took off from Governor's Island in New York Harbor in a general westerly direction. A few days later he cracked up his 50-hp Curtiss biplane at Addison, New York, 300 miles away, and quit the race.

Calbraith Rodgers had taken his license on August 7. With a Wright Model B that he had bought for $5,000, he then annexed $11,000 in prize money at Chicago's International Meet the same month. That was not enough, however, to cover the expenses of a trip across the continent; and "Cal" secured the backing of J. Ogden Armour, the nationally known meat packer—who was then promoting a grape soda drink called Vin Fiz. As part of the bargain, Rodgers agreed to christen his biplane the Vin Fiz Flyer, to paint the name on wings and tail, and to strew leaflets from the air advertising the product from one end of the country to the other. The difficulties of the flight were staggering, viewed in the light of facilities available today: no prepared landing fields, no regular supplies, no advance weather reports, no radio or other instruments, no flight plans. But the fame of Cal Rodgers was to endure a good deal longer than that of Vin Fiz.

Rodgers, 6 feet 4 inches tall, weighing 200 pounds, and a chronic cigar chomper, left the Sheepshead Bay racetrack, near New York, on September 17, wearing black leather boots, padded vest, black gauntlets, goggles, and a cap turned backward. He came down at Middletown that evening after flying 105 miles—a good start, but scarcely representative of what was to come. The mechanics on the five-car special train that followed the plane's route were never idle: landings were seldom made without damage, takeoffs were often disastrous; breakage of engine parts, the stresses and strains of bad weather, and the inroads of souvenir hunters required almost daily repairs. When the aviator received a charge of metal splinters in the arm from a broken connecting rod, it nearly ended the flight; the long delay that ensued was painful in more ways than one.

By October 10, when the Hearst offer ran out, Rodgers had beaten the existing distance records of Harry Nelson Atwood. Atwood, a serious-minded electrical engineer trained at the Massachusetts Institute of Technology, had abandoned his garage business near Boston to enroll in the Wright school at Dayton. He had quickly become a proficient flyer—"The Undisputed Eagle Of The Air," according to the headlines. Within three months of his first flight, he had flown from Boston to New London, Conn.; around a skyscraper in New York; and to Baltimore and then on to Washington—where he had landed on the White House lawn. Between August 14 and 25 he had also made a remarkable long-distance flight, with eleven stops, from St. Louis to New York: 1256 miles in 28 hours 58 minutes net flying time. Nearly all of his $10,000 prize, it was reported, went for expenses.

The patched-up Vin Fiz soon left Atwood's accomplishments far behind. Even so, Rodgers had lost the prize; but he obstinately refused to give up. On November 5, with only a few more miles to the Pacific, he landed, pelted with flowers, to a hero's welcome at Tournament Park in Pasadena. However, he was still not satisfied—Rodgers wanted to dip the landing gear of his machine in the water, to symbolize the linking of two oceans by air. But as he took off, the engine stopped, and the plane nose-dived into a ploughed field. The result was two broken legs, a broken collar-bone, and a brain concussion.

Eighty-four days after he had started, Rodgers—crutches lashed to his plane—finally achieved his goal at Long Beach. It had taken him 3 days 10 hours 4 minutes of flying time to cross the continent—for which the Aero Club of America awarded him its gold medal on December 13, 1911. "I expect to see the time when we shall be carrying passengers in flying machines from New York to the Pacific coast [at] an average of around a hundred miles an hour," he predicted. "That . . . cannot be done until some way is devised to box in the passengers against the wind." Cal did not live to see his vision fulfilled; but the well-known Wright biplane that exposed its pilot to the full force of the wind was already on its way out. With increasing speed, more protection was needed—and it was not long before passengers were "boxed in" by machines with a covered fuselage.

On April 3, 1912, not five hundred feet from where it had touched the Pacific waters, the Vin Fiz went out of control on an exhibition flight and plunged into the sea; Rodgers died of a broken neck and back. His plane is a national relic in the Smithsonian Institution today.

Robert Fowler tried again from Los Angeles on October 19, and finally made it over the mountains of southern California. In a series of alternate hops and mishaps, he reached the Florida coast at Jacksonville on February 8, 1912—the first to make the west-to-east crossing. On April 27, 1913, accompanied by a motion picture photographer, he flew over the Isthmus of Panama, from the Pacific to the Atlantic, in 57 minutes to register the first "nonstop transcontinental" flight—a feat which not only

In America's first airmail delivery, on September 23, 1911, pilot Earle Ovington carried a cargo of 1900 letters and postcards between Nassau Boulevard and Mineola, Long Island. (Author's collection)

surprised but alarmed Congress. The caper caused President Wilson to issue an executive order forbidding unauthorized flights over the canal and its defenses.

Headlines in the *New York Times* of Sunday, September 24, 1911, signaled one other newsworthy event in America: "Ovington Takes First U.S. Mail Through Air—10,000 People See Aviators In Novel Stunt At Nassau Boulevard Meet." As a feature of that suburban New York gathering, Earle Ovington, veteran of the Harvard and other meets, was sworn in as "air mail pilot number one"; and his Blériot-type Queen monoplane, powered by an Indian-brand rotary motor, became the first official mail carrier in America's skies. Postmaster General Frank H. Hitchcock handed Ovington a pouch containing a random selection of 640 letters and 1280 postcards, weighing about ten pounds. With the sack jammed between his knees, Ovington flew from "Aeroplane Station No. 1" to Mineola—a distance of 3 miles—and dropped his unique cargo at the feet of the waiting postmaster. The flight was performed as a "demonstration," not as the inauguration of a service, and the records state that it was made "at no expense to the government." A total of 32,415 postcards, 3,993 letters, and 1,062 circulars were similarly transported during that week; their cancellation marks are valued collector's items today.

As it happens, the Ovington flight was not the world's first aerial post. Hans Grade, the German pioneer, carried the first recorded piece of airmail—stamped by the post office *"Via Grade Flieger"*—in 1909, for the chamber of commerce of Bork, Germany. Nor did Ovington's flight antedate a historic air delivery by the Frenchman Henri Pequet in India on

February 18, 1911. The adventurous Pequet took a British Humber monoplane to the All India Exposition at Allahabad—in the course of which he flew locally a total of 6500 pieces of mail.

Shortly before the Nassau Boulevard meet, too, the first aerial post was inaugurated in the United Kingdom. To celebrate the coronation of King George V on June 22, 1911, special postcards were printed "for conveyance by aeroplane from London to Windsor." They carried the following warning: "No responsibility in respect of loss, damage, or delay is undertaken by the Postmaster General." C. H. Gresswell, who had learned to fly under the tutelage of Grahame-White in a British-built Farman at Brooklands, is credited with having carried the first pouch from the aerodrome at Hendon to an improvised flying field near Windsor Castle early in September 1911. Thousands of the postcards, which pictured a Henry Farman in flight over Windsor, were sent by enthusiastic patrons—including King George and Queen Mary.

It was an exciting new departure in the use of the aeroplane—but one that few believed would develop into the role taken for granted today.

CHAPTER SEVEN

Speed Is King

Why France had so rapidly forged into the lead, overtaking the United States and outflying all comers in the sky, is not difficult to explain in retrospect. First, Blériot was a Frenchman; his exploit in crossing the Channel gave France a sense of pride it had not known for years. To a nation that had lost the Franco-Prussian War in 1871, the Channel flight was a symbol of victory—almost literally a gift from heaven. Blériot's achievement was an emotional release for thousands of his countrymen; it caused women to weep and the blood of young men to run faster; it did more to restore the prestige of the nation than all the political debates and ideology of its statesmen.

Clément Ader, the distinguished French inventor and engineer, had remarked: "Whoever will be master of the sky will be master of the world." The aeroplane not only revived the people's spirits but hinted at greater glories to come. There followed a sort of holy fervor in the cause of flying, fed and kept alive by a patriotic press, subsidized by an equally patriotic array of French firms and individuals.

The stimulus given to aviation was coupled with a newfound devotion to sport. At the end of the nineteenth century, the Frenchman's indifference to physical activity was proverbial. Oscar Wilde (who spent the last years of his life in France), when asked what form of sport he indulged in, replied, "Dominoes, occasionally, on the terrace of a Paris café." That fairly reflected the attitude of most Frenchmen. But from 1905 on, it was a different story. The haunting possibility of another war sparked a vague desire for bodily fitness like that of the athletic Anglo-Saxons. The bicycle and the automobile became passions; tennis, rugby, football, boxing, swimming, even golf, began to take hold. As the swirl of world events began to exert change, it behooved France to get into shape; and sports as a way of life were impressed on the public consciousness by lavishly illustrated periodicals. Bicycle racers became national heroes—and aviators soon joined them. In a supplement to its issue of December 2, 1911, *La Vie au Grand Air* included color portraits, suitable for framing, of such airborne demigods as Védrines, Garros, and André Beaumont.

Another factor in the dynamic French approach to flying may have been the best-selling science fiction of Jules Verne. In the opening years of the twentieth century, his stories prodded the imagination of French youth and inspired an active interest in science and invention. They may even have been responsible for the birth of the early motor firms—Panhard-Levassor, Roger-Benz, Peugeot, and similar first comers in the field of applied mechanics. Finally, the Gallic temperament—daring, brash, even reckless—together with a naturally strong competitive instinct, may have influenced young Frenchmen to take to the air more readily and in greater numbers than did their contemporaries around the Continent and across the seas.

At any rate, by the beginning of 1912, all the important records were held by the French. Courage alone could not account for such complete dominance; the French also had more advanced equipment than their foreign rivals. American machines, for instance, were handicapped by the lack of a light, all-purpose motor such as the Gnôme. Furthermore, there were relatively few cash incentives in the United States, and certainly much less patriotic initiative than in France, to encourage research or competition.

The principal records for speed had been set by Edouard Nieuport six months previously and were still unbeaten. Nieuport had made good use of the discovery that a flatter camber permitted a correspondingly greater speed. On June 21, 1911, he had flown at Mourmelon at the rate of 133.13 km/hr. He had carried a passenger the same day at 108 km/hr; and previously, on March 9, he had flown with two passengers at 102.85 km/hr. All three records still stood at the opening of the new year. Even the American records for speed were held by foreigners: the 109.23 km/hr (67.87 m/hr) set by Leblanc at Belmont Park on October 29, 1910, remained unchallenged in the United States, as did Grahame-White's mark for pilot with passenger, made at the second Squantum meet on September 4, 1911. Nor had Sopwith's relatively slow 56.25 km/hr (34.96 m/hr) for pilot with two passengers, registered at Chicago with a British-built Wright, yet been surpassed by an American plane or pilot.

The world's distance records were owned by Armand Gobé, who had celebrated the day before Christmas, 1911, by flying 740 km nonstop in a 70-hp Gnôme-powered Nieuport at Pau; and by Emmanuel Hélen, who had traveled 1252.8 km at Etampes on September 8, with three stops. The best an American could show in the distance category was a flight of 283.62 km (176.23 miles), made by St. Croix Johnstone—who had taken his certificate on December 28, 1910, at the Blériot school in Hendon, England. Johnstone had established that mark on July 27, 1911, with a Moisant monoplane at Mineola, Long Island—a mark that proved to be his monument, for three weeks later he was killed at Chicago when his plane came apart in mid-air.

Meanwhile Georges Fourny, chief pilot for Maurice Farman at Buc, had set a world's record for duration that was to stand for nearly two and a half years. On September 11, 1911, Fourny flew without a stop for 13 hours 17 minutes 57 seconds in the firm's 70-hp Renault-motored biplane—an achievement that made the American record of 4 hours 16 minutes 25 seconds (set by Howard Warfield Gill in a Wright biplane) seem feeble by comparison. As for altitude, Roland Garros had soared to 4960 meters above the French seaside resort of Dinard on September 6, while America's best effort was that of Lincoln Beachey, who had climbed to 3527 meters (11,573 feet) in his 60-hp Curtiss during the 1911 meet at Chicago.

Only one world record worthy of note belonged to the United States on the first day of 1912. Thomas deWitt Milling, a U.S. Army lieutenant who was to become Chief of Staff of the Air Corps in World War I, had carried two passengers in a Burgess-Wright for 1 hour 54 minutes 42⅗ seconds at the Nassau Boulevard meet on September 26, 1911. But even that mark was shortly to succumb to a foreigner.

Not many Americans had in fact sought licenses. By the end of 1911 only 82 certificates had been granted by the Aero Club of America; even the British (whom Grahame-White was trying to shake from their lethargy by touring the country with a plane labeled "Wake Up England!") had obtained nearly twice as many—162. On the other hand, a number of United States citizens had acquired licenses abroad. Among them, for example, were Hayden Sands, J. A. Cummings, A. J. Houpert, Earle Ovington, Samuel Pierce, D. La Chapelle, Gardner Hubbard, William F. Whitehouse, James Lewis, R. H. Depew, and Edson F. Gallaudet, in France; and—in addition to St. Croix Johnstone—James V. Martin, Harry B. Brown, and W. M. Hilliard, in England. The zest for flying, and the coveted prestige of being an aviator, seemed far greater in Europe than in the country of the aeroplane's invention.

Instead of the well-organized competitions that flourished on the Continent and the professional technique displayed by rival firms and pilots that produced such highly crafted machines as the Nieuport and Deperdussin, barnstorming with the same old models was the order of the day in America. Instead of venturing into unexplored realms of engineering design, the Wrights and Curtiss adhered stubbornly to their pusher biplanes—a type headed for extinction. American monoplanes (with the notable exception of a bullet-nosed machine with shaft-driven propeller at the tail, built by Edson Gallaudet, an engineer from Connecticut) were more or less obvious copies of European prize winners. The best example was an Anzani-powered craft, with rounded wing tips and wing warping for lateral control, built by Albert and Arthur Heinrich of Baldwin, Long Island. This machine combined characteristics of the Blériot and Morane-Saulnier.

Instead of spending money developing the flying machine, Americans concentrated on the mass production of motorcars. The Ford, Maxwell, Reo, Chevrolet, and Hupmobile were placed on the market at less than a thousand dollars; the Cadillac touring car cost about twice as much; and the elegant Mercer sold for $2600. Purchase of an aeroplane, by contrast, demanded an outlay of well over $5000. Most of those who gave flying any attention at all were content to watch the exhibitions of migrant bird-men who, daring and dashing as they might be, were often little more than mountebanks. As a circus attraction, as a climax to conventions or other outdoor festivities, the aeroplane was booked as a novelty to entertain the customers or to provide a thrill for those who were willing to pay a few dollars for a ride. In its issue of August 17, 1912, *Aero & Hydro,* a weekly devoted to flight, reported that the Curtiss pilot Beck-with Havens had competed for the attention of spectators at Ontario Beach (near Rochester, New York) with ten polar bears, a clown, band music, and fireworks. Among other items of news, Charles F. Walsh, an-other Curtiss aviator, had performed at the Clark County Fair in Spring-field, Ohio; Paul Peck had carried passengers in flight for the Merchants' Association at Brazil, Indiana; and Jimmy Ward was exhibiting at Du-luth under the auspices of the Elks. At Grand Forks, North Dakota, Dr. Frank Bell, taking up a parachutist, had had a bad fall from fifty feet when (according to an eyewitness) one of the control wires had become "deranged."

Under the heading "Among The Aviators," a gossipy column in the magazine shed a good deal of light on the state of flying in the United States, just then at its seasonal peak. George Schmitt, with a Baldwin-type biplane, gave a successful 20-minute demonstration at Bellefontaine, Ohio. Charles K. Hamilton "badly damaged" his machine in a field that was too small at Hartford. Charles J. Hibbard made two flights at Miami, Ohio, after the second of which he landed in a cornfield, breaking an arm and "badly damaging" his machine. George Underwood, using a Sparling bi-plane with a Kirkham engine, barnstormed in Missouri from Kirksville to Glenwood and thence to Queen City, stopping off at Greentop to greet Ed Korn, who was giving exhibitions at that town with his Benoist tractor.

Still other aerial doings were recorded. Earl Dougherty, testing the tractor biplane with 80-hp Hall-Scott motor and upturned wing tips built by William E. Sommerville, the mayor of Coal City, Illinois, reached a height of 2000 feet. Grover Cleveland Bergdoll, millionaire Philadelphia sportsman, obtained a barograph reading of 6200 feet in his Wright pusher at Llanerch, Pennsylvania. Walter E. Johnson tried out a Curtiss hydroaeroplane on the surface of Lake Seneca at Geneva, New York; and Harry B. Brown made news by flying with a passenger in a "gale" for 57 minutes near New York.

Unless a flyer met with an accident, fatal or otherwise, the curiosity generated by these small-town exhibitions was little different from that aroused by any outlandish phenomenon—such as Siamese twins, a bearded woman, or the imperishable young man on a flying trapeze. When F. W. Kemper exhibited in his Wright machine at Silver Lake, near Akron, the flights were successful; but the people of the district "did not appear to be deeply impressed with aviation, as the attendance was very small." At the same time, when Art B. Smith, piloting a Mills biplane, treated the denizens of Deadwood City, South Dakota, to their first sight of a plane flying over the Black Hills, he was presented with a gold medal to celebrate the occasion. The fact that the gold in the medal came from the aforesaid hills seems to have been of more significance to civic-minded officials than the flight itself.

Women, too, went on the exhibition circuit, and their sex lent added attraction to the shows. As well known as that of any man were the names of Matilde Moisant, whose brothers pioneered the barnstorming tour, and of Harriet Quimby, her frequent teammate and the first woman in America to receive a pilot's license. The first solo flight by a woman in America is credited to Blanche Stuart Scott—described as "an attractive and well-built aviatrix"—of Rochester, New York, made on a Curtiss-type machine on September 2, 1910. Bessica Raîche, whose husband François was the builder of an experimental biplane for the Aeronautical Society of New York, is also reported to have flown in September 1910. Others who became licensed pilots and, especially after 1914, achieved a permanent place in the records of the pioneers were Katherine Stinson, who used a Wright biplane and motor to gain a certificate on July 24, 1912; and Ruth Law, who took hers on November 12, 1912, with a Curtiss biplane and motor. Both of these were star exhibition flyers. Florence Seidell, a Curtiss pilot; Mrs. Richard Hornsby; and Marjorie Stinson, the second member of the family to graduate from the Wright school, received licenses in 1913.

Matilde Moisant was granted a certificate on August 13, 1911—naturally enough, in a Gnôme-powered Moisant monoplane of the familiar Blériot type—the second woman in America licensed to fly. At a time when the hairbreadth escape was a part of every other flight, she was renowned for her frequent flirtations with death—particularly while barnstorming through the United States and Mexico in 1911 and 1912 as a member of the team of Moisant and Quimby. The first woman to be awarded an altitude prize in America—for some 2500 feet at the Nassau Boulevard meet in 1911—Matilde was an inveterate believer in the number "13." Her birth date had been a Friday the 13th; she won her license on the thirteenth day of the month; and her machines were always numbered "13." A victim of four serious crashes, she stopped flying temporarily after being dragged from her burning plane in Texas, clothing aflame; a leak in the fuel tank had caused a fire after a hard landing. But Matilde Moisant lived longer than most of her contemporaries, male or female. She died in Los Angeles, at the age of 85, in 1964.

Harriet Quimby took her license at the Moisant school a couple of weeks before her companion Matilde—and made news by reportedly flying her plane at the rate of a mile a minute. She rapidly gained more experience, both in the United States and in a series of exhibitions in Mexico. On April 16, 1912, she acquired lasting fame by becoming the first woman to cross the English Channel, flying through fog from Dover to the beach at Hardelot, France, in a Blériot she was using for the first time. Before taking off on that trip, she had been instructed in the use of a compass by the more experienced Gustav Hamel. So anxious was he for her success that he offered to don the distinctive mauve satin flying costume she affected, make the flight, land at an unfrequented spot, and give her the credit. "Be sure to keep your course, whatever you do," he told her, "for if you get five miles out of the way, you will be over the North Sea, and you know what that means." Hamel might have taken to heart his own words a couple of years later, when he himself flew off into the Channel mists and was lost forever.

Although a cool, careful, and capable pilot, Harriet Quimby did not strap herself in when she took part in the 1912 meet at Boston's Squantum flying field. A few weeks before leaving for Boston, she had written an article for *Good Housekeeping,* entitled "Aviation As A Feminine Sport," which aimed at giving women confidence in their ability to equal the performance of men. "There is no reason to be afraid," she had written, "as long as one is careful. . . . Only a cautious person . . . should fly. I never mount my machine until every wire and screw has been tested. I have never had an accident in the air." On July 1, Miss Quimby, together with her manager, W. A. P. Willard, was rounding Boston Light in her Blériot when the monoplane was caught in turbulent air. Both pilot and passenger were flung out, plummeting over a thousand feet to their deaths in the sea.

Three other women had preceded Harriet Quimby in death while flying. The tragic distinction of being the first belonged to Denise Moore, an American who crashed to earth at the Henry Farman school at Etampes in the summer of 1911 while attempting to qualify for a license. The second fatality occurred near the same spot on March 10, 1912, when Suzanne Bernard, a nineteen-year-old French girl, was killed while completing her tests for a pilot's license. In making a turn, her Caudron biplane was caught in a wind gust and capsized, falling two hundred feet and crushing the young aviatrix beneath the wreckage. The third woman flyer to die also was an American—Mrs. Julia Clark, who crashed at Springfield, Illinois, shortly after taking her license in a Curtiss biplane on May 19, 1912.

Of the numerous stars of American exhibition flying, only one shone with brilliance in the area of design and construction as well. He was Glenn Luther Martin—self-taught pioneer with a Curtiss-type pusher, builder of the Martin tractor, and one of the earliest experimenters with hydroaeroplanes and flying boats.

Born at Macksburg, Iowa, on January 17, 1886, Martin spent his boyhood in Liberal, Kansas, where his father ran a wheat farm and hardware store with indifferent success. Like other pioneers, young Glenn flew kites and worked in a bicycle shop; and when the family moved to Salina, he became a helper in a garage. To escape the hard midwestern winters, the family pushed on to California in 1905. At Santa Ana, where he went into

Glenn Luther Martin, a self-taught aviator, delivered copies of the Fresno Republican *by parachute from his makeshift biplane. A ready publicity seeker, Martin claimed to be the first flyer to take his mother aloft, to shoot motion pictures from the air, and to bombard a make-believe fort with flour sacks. He starred opposite Mary Pickford in a 1915 Hollywood film. (National Archives)*

the business of selling and repairing automobiles, Glenn heard about the Wrights. Determined to fly, he built and practiced with a biplane glider. With maternal (though not paternal) approval, Martin rented an abandoned church for twelve dollars a month and, by the light of a kerosene lamp, began the construction of a powered machine.

In 1909 his makeshift biplane of bamboo and varnished cotton cloth, powered by a secondhand Ford motor of 15 hp, emerged from its cocoon and made a few hops, skips, and jumps around a nearby mesa. "For Heaven's sake, if you have any influence with that Wild-eyed, Hallucinated, Visionary young man, call him off before he is killed," begged the family physician, Dr. H. H. Sutherland, in a letter to Mrs. Martin dated December 30, 1910. But by that time Glenn, an indefatigible member of the trial-and-error, do-it-yourself school, was giving public demonstrations, breaking wood and taking tumbles in his stride. The following year he was in demand at Fourth of July celebrations and county fairs; and by 1912 he had established himself on a professional basis as a daredevil showman. On January 20, 1912, the enterprising aviator took off with a sack of mail from the Dominguez field and, shortly thereafter, dropped it into the outstretched arms of the waiting postmaster at nearby Compton. "Pacific Aerial Delivery Route Number 1 Opened By Glenn Martin" was the front-page story in the Los Angeles *Tribune* next day.

At Chicago's International Meet that summer, Martin began to collect a scattering of the prize money then to be made in United States aviation—$4854, to be exact. It was not much compared with the big sums to be won in the European races; but it encouraged him to move his "factory" to Los Angeles. There he not only produced more planes but opened a training school for flyers.

Highly sensitive to the value of publicity, Martin made more headlines by chasing coyotes in his biplane; by dropping a baseball into a catcher's mitt or a bouquet into a May Queen's lap; by hunting escaped convicts or searching for aviators missing over the ocean; by releasing a blizzard of department-store advertising on a goggle-eyed public. Martin claimed to be the first flyer to take his mother aloft, the first to shoot motion pictures from a plane, and the first to bombard a make-believe fort—with bags of flour—from the air. Although cast by nature as an aloof personality, with thick-lensed glasses that belied his daring in the air, Glenn Martin achieved stardom in a Hollywood film with Mary Pickford in 1915; a stern, sober-minded, unheroic type, he played a romantic (and out-of-character) role—as an aviator—in *The Girl of Yesterday*.

While American aviation maintained its indifferent stance, featuring desultory flying shows, solo exhibitions, and stunting, Lord Northcliffe and his *Daily Mail* sponsored the first "aerial derby"—the Circuit of London—on June 8, 1912. The newspaper had begun to promote aerial tours around the country in May, and the derby got the English season off to a highly publicized start. Not only did the event contribute mightily to public interest in flying, but it gave the British an idea of what some of their own machines and pilots could do. If aviation was to develop along practical, useful lines, so reasoned men like Northcliffe, a prize competition for cross-country flight was one way of testing planes with an eye to producing better ones in the future.

Starting at one-minute intervals from Hendon, contestants in the 81-mile race took off in the following order:

	Pilot	*Machine*
1.	S. V. Sippe	Hanriot monoplane, 50-hp Gnôme
2.	T. O. M. Sopwith	Blériot monoplane, 70-hp Gnôme
3.	G. Hamel & passenger	Blériot monoplane, 70-hp Gnôme
4.	P. Verrier & passenger	M. Farman biplane, 70-hp Renault
5.	W. B. R. Moorhouse	Radley-Moorhouse monoplane, 50-hp Gnôme
6.	E. Guillaux	Caudron monoplane, 45-hp Anzani
7.	J. Valentine	Bristol monoplane, 50-hp Gnôme

Six other entrants, for one mechanical reason or another, were left at the post. But the popular Gustav Hamel, who had helped to ferry the first

English airmail between Hendon and Windsor, won fame for himself and for an adventurous young lady named Trehawke Davies—by carrying her as a passenger in his Blériot. It was a day of adventure for others too: some lost their way, some encountered bad weather, some dropped out owing to engine trouble. Tom Sopwith—later to gain fame as a designer-constructor of both land- and sea-based planes—was disqualified on a technicality, but appealed the judges' decision. Five months later the Royal Aero Club allowed his appeal and declared him the victor. It was not the first (nor the last) important honor Sopwith gleaned: in 1911 he had won the £4000 Baron de Forest prize for the longest flight in a British-built machine from England into the Continent.

Although five of the seven machines taking part, including Sopwith's, were French, it was more than a consolation to the British that second and third places were taken by homemade products. The official results, prior to Sopwith's appeal, were as follows:

Pilot	Time	Prize
1. Hamel	98 min 46 sec	£250
2. Moorhouse	120 min 22 sec	£100
3. Valentine	146 min 39 sec	£50

In the middle of June, France held a two-day competition that for the first time focused attention on what aviation might mean to the military mind. The success of the first town-to-town race, from Angers to Saumur, in 1910 had inspired René Gasnier and his brother Pierre to organize a more pretentious competition. The Circuit of Anjou was the result: a triangular course of 157 km between the towns of Angers, Cholet, and Saumur—to be run three times the first day, four times the next. It fanned a latent French army interest in the capabilities of the aeroplane in two categories: scouting and bombing. Instead of being strictly a sporting event, it was the first contest intended to show to what extent the aeroplane could be depended upon in war.

A surprising total of thirty-five constructors entered their machines. Among the planes taking part were the Blériot, Deperdussin, Morane, Hanriot, R.E.P., Nieuport, and Borel monoplanes, and the Henry Farman, Bréguet, Astra, Caudron, Sommer, Zodiac, Zens, and Ladougne (La Colombe) biplanes. It was to be a test of speedy transportation of a useful load: points were to be awarded on the basis of actual speed multiplied by a coefficient calculated in accordance with the number of passengers carried. Most of the military attachés accredited to Paris were in attendance, keenly interested in the potentialities of the different models. The French government was represented by Finance Minister Besnard—a first indication of the place that aviation would soon occupy in the budget. Prize money amounted to 120,000 francs.

Excitement rose as the dates of June 16 and 17 drew near. From every corner of France and from many points abroad an enormous crowd had gathered. At the starting point in Angers, the meanest hotel commanded a fabulous price for its rooms; in the streets and cafés swarmed a keyed-up, jostling throng; the strident cries of hawkers, with banners and streamers flapping overhead, contributed to the impression of a mammoth carnival.

Practically all the famous pilots of Europe were on hand, as well as others not yet nationally known. A few of the names ticked off on the scorecard were Védrines, Garros, Brindejonc des Moulinais, Espanet, Bielovucic, Gaubert, Renaux, Bobba, Bathiat, Legagneux, Gobé, Tabuteau, Hélen, Bedel, Vidart, Prévost, Perreyon, and a lone entrant from across the Channel—Gustav Hamel. On the eve of the race, the press published betting odds on France's own great aerial "derby." Odds-on favorite was the slick, smooth Deperdussin, quoted at 1 to 4, with a high-powered team composed of Védrines, Vidart, and Prévost. The Nieuport was quoted at 3 to 1, the Morane at 4 to 1, the Blériot at 6 to 1, the Bréguet and Farman at 5 to 1. The backing given to the newest-model Deperdussin seemed well founded. An elaborate display arranged by the personable Armand Deperdussin (aided by a large public-relations staff occupying an entire floor of the leading hotel) presented the plane to the public in a capacious hangar.

To the dismay of promoters, public, and pilots alike, the dawn of June 16 broke in a violent storm, with sharp showers of rain, black clouds scudding overhead, and a wind that bent the poplar trees almost double. Arguments raged like the gale itself as to whether the race should be allowed to start. There were few stormy petrels among the birdmen. Védrines, in good socialist style, mounted a tin of gasoline and harangued his colleagues on the risks of taking off; but his call for a pilots' "strike" fell on at least one pair of deaf ears. At nine o'clock, wearing an old jersey and white woolen headgear, Garros brought out his Blériot and asked the committee's permission to take off. Wind whistled through the bracing of the fuselage as four mechanics held down the machine. With the deepest misgivings, the judges gave their assent. In seconds the Blériot was airborne, flying this way and that until its pilot managed to gain enough altitude to clear the trees and head toward the leaden clouds.

Roland Garros, despite his reputation as a second finisher, was to come into his own that tempestuous day. By training and experience he was well qualified for the moment. During his student days, he had found himself intrigued by the advent of the motor age; and subsequently he became an agent for a French-made roadster. Later, as a spellbound onlooker at Rheims in 1909, he decided that flying was to be his career. A Demoiselle built by the Clément-Bayard factory for Santos-Dumont was at that point the least expensive machine on the market. Garros bought the capricious, frail "butterfly" of bamboo and silk for 7500 francs, and with other self-taught amateurs made his headquarters in the cavernous dirigible hangar at Issy-les-Moulineaux. After gradually learning how to get off the ground and cautiously execute turns, Garros embarked with a group of equally

Roland Garros (at left), winner of the 1912 Circuit of Anjou. Under hostile skies Garros navigated by compass, and even stopped to ask a farmer directions. For his first-place finish, the Aéro Club de France awarded him 75,000 francs. (Photographed by the author.)

inept performers on a series of small-town contests and exhibitions—
Cholet, Rennes, St.-Malo, Dinard, Evreux, and La Ferté–Vidame.

From the viewpoint of man's progress in emulating the birds, these
shows left much to be desired—for they consisted largely of hops and
crashes. But by this clumsy and erratic process Garros acquired the nerve
to sign up at Belmont Park as a replacement for the injured Léon Morane, whose near-tragic accident had occurred a couple of weeks before.
When he returned to France after his American tour with the Moisant flying circus, Garros was on the threshold of success—adept in the art of
taking off from seemingly impossible places, in flying through wind and
weather to satisfy unruly crowds, and in making tight twists and turns designed to thrill but also to serve a useful purpose. For instance, he had developed the spectacular *tire-bouchon,* or corkscrew descent, from a great
height with motor cut off, which enabled him to pinpoint his landing spot
in case of motor failure. "He was like a bird, turned into a man," declared Jacques Mortane, the aviation writer, early in 1912; "the air was
his element . . . it was impossible to describe the impression made by his
evolutions . . . nor was he reckless, for he knew exactly what he could
demand of his plane."

Exactly the opposite in approach and disposition from Védrines, his
principal rival, Garros was a rather debonair figure who kept aloof from
the crowd yet had a congenial personality. His rise to fame really began
with a spur-of-the-moment decision to enter the Paris–Madrid race in
1911. A year later he put in for a starting position in the Circuit of Anjou, having just returned from an expedition to South America with three
veteran comrades of the Moisant days: Audemars, Simon, and Barrier.
The trip of these four musketeers to the Southern Hemisphere had been
financed by an "angel" providentially encountered in the United States—
Willis McCornick, under whose benevolent eye the Queen Aviation Company Ltd. had been formed to regale the masses in Brazil and Argentina.
The experience had given Garros even more skill, sagacity, and *sang-
froid*. Still lacking was experience in cross-country flying, except for his
attempt in the Paris–Madrid race; but Garros was no worse off in that
respect than many others.

Pitted against more powerful machines, and under normal conditions,
the light Blériot of Garros would have had little chance at Angers. But
the pilot's refusal to be intimidated by the elements gave him an advantage—for less than a dozen others dared to follow his example. With no
windshield, the rain beat fiercely in his face and obscured vision through
his goggles. Hail and fog also beset him. It was "horrible," Garros acknowledged afterward: "I controlled the plane with one hand and shielded
myself with the other. I could see absolutely nothing and navigated solely
by compass. Finally, I saw a hole in the clouds, landed in a small field, and
asked a farmer where I was. He gave me the right directions and I took
off again."

Only one other antagonist threatened his lead: Brindejonc des Mouli-

Commemorative stamp issued for the Circuit of Anjou. According to the press, the meet demonstrated that France had "machines capable of flying in all kinds of weather."

nais, a nineteen-year-old representative of the Morane-Saulnier camp, who with tears in his eyes had begged his principals to let him brave the storm in pursuit of Garros. With a 70-hp motor (Garros's was of 50 hp), he was close behind; but for crossing the finish line 7 minutes after the race closed at 6:30 P.M. he was disqualified. Others dropped out because of the strain to which the rough air subjected them and their machines; some were simply too airsick to continue.

The performance of Garros on that first day of the meet—and, under less hostile skies, the second day as well—was amply rewarded. He completed a distance of 1101 km in 15 hours 40 minutes 7 seconds flying time, taking the grand prize of 50,000 francs offered by the Aéro Club de France; a first prize of 20,000 francs for speed; and a prize of 5,000 francs for fifth place in another special event.

Mortane's praise of Garros was justified. In a biographical sketch introducing the short *Guide de l'Aviateur,* which the flyer published in 1913, Mortane pointed out that hard luck had at last given way to the "most marvelous show of valor in aviation," and that his exploit against a fierce wind placed Garros "ahead of all the pilots in the world."

In his recollections of the Angers scene, Pierre Gasnier tells of two incidents on the lighter side. The thousands who had gotten up for the early start had to do without breakfast. Discovering a well-stocked luncheon buffet of chickens and lobsters around noon, the famished crowd began a raid that became an uncontrolled riot. Seeing the victuals disappear, to his imminent financial ruin, the caterer rushed for a pistol and melodramatically shot himself—although, as it turned out, not fatally. Later in the meet, an Aerobus, one of the larger planes of the period, piloted by a Voisin flyer named Maurice Allard, flipped over and plummeted 150 meters, burying four passengers beneath the wreckage. All walked away; but, to the uproarious laughter of the crowd, a beautiful blonde—the girl friend of the pilot—emerged with a bare skull, the golden tresses of her wig impaled on a broken strut.

France had scored again with the Circuit of Anjou. "The results proved for the first time that we have machines capable of flying in all kinds of weather," crowed *Le Matin* next morning. "It was a moving, at times agonizing, demonstration." Said the *Petit Journal:* ". . . our aviators have shown, for the first time, that the French army can count on them."

It was also the last time that the Blériot won a decisive victory. The machine that had conquered the Channel was already being outmoded by newer, swifter, and sturdier types. Moreover, the number of accidents to which it was subject seemed on the increase, to the concern of high officials in both France and England. In March 1912 the French army temporarily suspended the use of all monoplanes following a report by the Blériot firm that discussed weakness in the top bracing. With the exception of Levavasseur's advanced low-wing Antoinette with streamlined landing gear (specially designed and built for the army trials but too heavy to fly), the modern cantilever principle of monoplane wing construction was still unrealized. Wing failures were in fact becoming so frequent that in September of the same year a provisional ban on all army monoplanes was also ordered in England, pending the report of a committee appointed to investigate the many crashes.

Garros, however, would not turn his faithful mount out to pasture. It was his belief that the disasters that had befallen the Blériot were due partly to overweight, resulting from attempts to reinforce vital parts, and partly to faulty maintenance—rather than to any inherent defect in design. Out of loyalty to the machine that had brought him renown, he presented his 20,000-franc aeroplane to the army, in the hope that its successful operation would absolve the other Blériots of blame. It was a futile gesture: by the time the offer was accepted, the French authorities were already about to turn to entirely new models.

The Circuit of Anjou was followed immediately by one of the most picturesque gatherings in Europe—a meet in Vienna, the capital of Austria-Hungary. In that part of the world, the imaginative designs of an engineer named K. Illner and of A. Warchalowski, holder of Austria's pilot license No. 1, had been among the most promising. Another Warchalowski—Karl—had produced a biplane on Maurice Farman lines, but with differently shaped ailerons and with the corners of the leading edge rounded. But it was Igo Etrich who, as early as November 29, 1906, made the first fledgling hop at Wiener Neustadt with the forerunner of one of the world's most successful planes. Etrich's Taube ("Dove") was a graceful, birdlike monoplane whose wing tips curved upward, giving great automatic stability; the motor was a 60-hp Austrian Daimler. The model proved so efficient that in 1911 the German Rumpler factory purchased Etrich's design outright and began producing Taubes of its own.

Postcard showing the aerodrome of Buc. (Photographed from a Maurice Farman plane by Lt. Arthur Noé.)

Georges Fourny, pilot for the Maurice Farman school at Buc and holder of the duration record in 1912. (Photographed by the author.)

Eugène Renaux, chief pilot for Maurice Farman and distance-record holder, in a Farman biplane at Buc, July 1912. (Photographed by the author.)

At the Vienna meet, on the historic plain of Wagram near Aspern, the Austro-Hungarian hosts outdid themselves with an intelligent organization of events, excellent facilities for repair, ample space for spectators, and unstinting hospitality. The French participants—among them Charles Nieuport, with two types of his brother's monoplane; R. Bedel, with a Morane powered by a 60-hp Anzani motor; and Garros, with his familiar 50-hp Gnôme-powered Blériot—were given plenty of competition. Garros was not able to repeat his recent success and finished only fifth in the 100-km race from Vienna to Neustadt and back.

The star performer among the machines was the Austrian Lohner—a two-seater biplane with staggered and partially V-shaped wings, fitted with a 125-hp Daimler motor, which proved particularly effective in the climbing and weight-carrying classes. On June 28, 1912, this striking machine, piloted by Z. R. Blaschke—a young lieutenant in the Austrian army—set a new world mark for altitude with passenger: 4530 meters. The French record holders could scarcely believe the figures; but at the invitation of the judges, Garros verified that the Austrian barographs were in perfect working order by taking them up himself to 1000 meters. First prizes for landing in the smallest circle and for accurate "bomb dropping" were taken by the Rumanian Ouvert Vlaicu in a strange-looking monoplane with rudder at the front, a chain-driven propeller at either end of the wing, a triangular tail, and a 50-hp Gnôme motor. This refreshing oddity was nicknamed La Folle Mouche ("The Crazy Fly"). The original model was purchased by the Rumanian army; but the antics of its successor were unhappily terminated a few months later by a fall that killed the inventor.

At right, a Blériot monoplane, knocked down for transport, is hauled by horses. At far right, a disassembled Deperdussin is pushed up a wooden ramp at Issy-les-Moulineaux. (Photographed by the author.)

At Vienna was germinated the idea for a spectacle that came to full flower in the second Austrian meet, held in 1913. Like jockeys in a horse race, the aviators lined up their machines at the starting line for the qualifying trials. On signal, the whole pack—monoplanes and biplanes of all shapes and sizes—moved down the field to take to the air together, regardless of which way the wind was blowing, in a bedlam of noise and a haze of blue smoke. It was a minor miracle that no machines swerved in the air or tangled on the ground. At the 1913 meet, however, a collision after takeoff brought severe injuries to Henri Molla, top pilot of the R.E.P. firm, and two Austrians in a Lohner.

As the year wore on, it was apparent that the aeroplanes of 1912 were no longer the shaky craft of the first few years of flight. The newer models showed more ruggedness of construction, a better application of the principles of aerodynamics, and, of course, more power. Body frames for monoplanes were less often of spruce or other light wood, and more usually of thin, high-alloy steel tubing. Steel joints were also beginning to appear. The conventional biplane design of the Wrights, with propeller and motor in the rear, was giving way to the tractor fuselage—a design that Americans were slower to adopt than Europeans. There were two schools of thought on the subject: if the propeller was in front, it worked in undisturbed air but threw its wake against the wings; if in the rear, it worked in air made turbulent by passage of the plane surfaces, but the plane had calm air ahead to enter. It is a question whether, had not Wilbur Wright died of typhoid on May 30, 1912, the new form would have

found favor with his firm more quickly; for it was not until after the start of World War I in Europe that a Wright biplane in tractor form was finally put on the market. This was the Model L, which had a covered fuselage similar to those of other contemporary conventional types.

Although biplanes were supposed to be more stable and structurally stronger, monoplanes were rapidly coming into their own. With monoplanes it was found that relatively greater lift could be obtained because one wing alone did not create interference in the air—a point that seemed to be more readily appreciated in Europe than in the United States. Monoplanes, moreover, offered less head resistance because they did not have the struts and wires of a biplane and therefore could slip through the air faster. According to a private computation, there were in 1912 nearly 200 different makes of aeroplane—beginning alphabetically with the Albatros and ending with the Zodiac—of which 100 were biplanes and 91 monoplanes.

Speed was, in fact, king in 1912—in the air as well as on land and sea. Aeroplane speeds of a mile a minute no longer caused astonishment. Now, with the lessons of lift and control more fully grasped, the problem of power became paramount—providing one of the main incentives for augmenting speed capability. Although engines sometimes had a disconcerting way of falling out while in flight, if the plane structure was too weak, they were certainly becoming more efficient and delivering more horsepower in relation to weight. For improved performance in the air, more speed was essential; an increase in power, in turn, meant more speed.

Nothing could better illustrate the gap between the scientific, technological approach of the French and the sluggish progress of American designers than the fourth Gordon Bennett race, held on September 10, 1912, at Chicago. The victory of Weymann at Eastchurch the previous year had brought the competition back to the United States; and hopes were raised in America that the pendulum was about to swing toward that side of the Atlantic once more. When the lists were closed on March 12, 1912, teams had been entered from Belgium, England, France, Holland, and Switzerland. It was thought that America, as the defending country, would put forth superhuman efforts to keep the cup, even if it meant importing a foreign machine for the purpose.

At first, doubt was voiced whether any craft from overseas would be able to compete at Chicago, because of the possibility that the Wright company, alert for infringement of its patent rights, would be tempted to institute court action. An injunction had been obtained against Grahame-White a short time before, forbidding him to use his plane in the United States; but Wright officials magnanimously assured the Aero Club of America that they would refrain from legal proceedings in the case of entrants in the international contest. With that threat out of the way, the organizers were optimistic. The expectation was that additional awards (with a consequent increase in the number of competitors) would be put up because of the greater distance—200 km (124.3 miles)—this year, compared with 150 km the year before.

America's chances dwindled, however, as the day of the contest drew near. Machines constructed in the United States had never approached the speeds obtained abroad, and money had never been forthcoming for new models to dispute French supremacy. At the last moment, however, the Chicago Cup Defender syndicate was formed. Through Norman Prince, a well-to-do sports enthusiast and himself a licensed pilot of the Burgess-Wright biplane, an urgent order was placed with the Burgess firm at Marblehead, Massachusetts, to build a monoplane for the competition. But when it was ready, nobody could be found to fly it. After a frantic search, Glenn Martin was brought from the West Coast as pilot, even though he had never flown a single-surface machine and had never traveled faster than a putative 70 m/hr in his biplanes. Martin insisted on alterations in the design, in order that the machine might in some measure conform with those of his own experience. As the hours ticked by, endless arguments brought more heat than light to the problems of structural changes.

Meanwhile a second American machine, the Columbia, was entered through the energetic efforts of Paul Peck, a twenty-two-year-old flyer from Charleston, West Virginia, who at the moment was holder of the American duration record. Peck had recently been suspended by the Aero Club of America for a technical violation of its rules; but with no alterna-

Paul Peck, the American entrant in the 1912 Gordon Bennett race at Chicago. His machine, the Columbia, never got off the ground because of a missing magneto part. Peck was killed in a crash shortly after the race. (Author's collection)

tive candidate available, he was reinstated at the eleventh hour in the face of the national emergency. Although the specially constructed Columbia biplane had only a 50-hp Gyro motor, the machine was considered fast by American standards.

On the eve of the event, all entrants from abroad had dropped out with the exception of the French. Glenn Curtiss, in Paris, looked back on his 1909 truimph and predicted flatly that France would win. "All America has is brand new machines and no one to fly them," he pointed out, "whereas France has the experience, the machines, and the men." It was enough to glance at the world records to confirm that France had the experience. France had the machines, too—in particular the clipped-wing, ring-cowled, streamlined Deperdussin *monocoque* racer, evolved by the engineer Louis Béchereau. (The designation *monocoque* was applied to a single-seater, single-engine, single-surface plane with cylindrical fuselage constructed of lengths of plywood and cloth wound spirally over a tapering frame. The result was a strong, lightweight, streamlined body. The term "ring-cowled" described the device of a circular metal ring around a rotary motor to prevent it from splattering oil.)

France had the men as well; and for this contest it chose its best—two

Deperdussin pilots, the redoubtable Jules Védrines and the versatile Marcel Prévost. The latter, in addition to flying this machine over land and sea, had made a world's altitude record for pilot and passenger (3000 meters in 1 hour 2 minutes) on December 2, 1911. The third member of the team was the brilliant André Frey, with a specially groomed Hanriot monoplane.

September 9 dawned blistering hot. For the first time in the history of the competition, contrary to expectations, no one had put up the money for prizes. Spectator interest was embarrasingly low. Fewer than 1500 people made the trek to the Chicago suburb of Clearing—a spot hard to reach by trolley, and a tiresome journey by auto or chartered train. By 9:20 in the morning, Védrines and his racer were in the air, circling the course at low altitude in masterful, if monotonous, routine, to become the winner at an average speed of 105½ m/hr. In the afternoon Prévost repeated the clocklike performance of his teammate, rounding the pylons only twenty or thirty feet above the ground, and coming in a close second. Frey withdrew his Hanriot after completing half the course when the motor showed signs of overheating in the broiling sun.

Americans waited in vain for one of the United States entrants to start. To the intense chagrin of the faithful who had clung to the hope that at least a token flight could be made, neither American machine was able to take to the air. The Burgess was still untested; and the loss of a vital magneto part that could not be replaced on brief notice prevented the Columbia from taking off. Jacques Schneider, a prominent, well-to-do supporter of French aviation, gave expression to the feelings of Frenchmen everywhere when he rushed impulsively onto the field to embrace Védrines after his landing and to drape the Tricolor around his neck.

The American failure was compounded by tragedy on the opening day of a meet at the Cicero field, near Chicago, immediately after the running of the Coupe Internationale d'Aviation. Paul Peck, testing the Columbia, was instantly killed when the machine went into a tight spin and fell—completely out of control—from a height of 1000 feet. Four days later, the 1912 Chicago meet was further marred: Howard W. Gill, an old hand with the Wright biplane, was mortally injured in a collision with George Mestach's Morane-Borel, when the landing gear of Mestach's monoplane tore away the tail assembly of the Wright. The accident lent credence to bitter charges by aviation enthusiasts that the meet had been mismanaged; that the officials were ignorant of the elementary rules of flying; and that the schedule of events had no scientific or sporting value whatever.

CHAPTER EIGHT

Over the Waves

In the pursuit of successful flight over water, Americans found an endeavor in which they could, at last, take equal honors with the French. To Glenn Curtiss belongs the credit for being the first to build and market a machine capable of landing on and taking off from the surface of the sea.

Yet the premier success of Curtiss was itself preceded by the experiments of a Frenchman little known today—Henri Fabre. After the futile experiments of Voisin and Blériot on the river Seine in 1905 and 1906, French aviation—indeed aviation everywhere—developed literally from the ground up, instead of from the water. Fabre, however, an inventive marine engineer and navigator, persisted in his original research on the problem of achieving powered flight from a water base. On the Gulf of Fos at Martigues, near Marseilles, he began a series of tests in 1909 with a weird contrivance he had put together during the course of his studies the year before. Resembling a giant dragonfly flying backward, it consisted of a skeleton framework mounted on three scientifically designed floats, with a 50-hp Gnôme motor and Chauvière propeller at the rear. On March 28, 1910, Fabre managed to lift his creation from the surface of the sea for the first time. He continued his flights, droning fitfully over the waves for short distances, until May—when the apparatus suddenly took a header into the Mediterranean and was almost totally wrecked. Fabre himself was unhurt. The machine reappeared at Monaco in 1911, during a series of aquatic races at that port; but the engine proved inadequate, and development proceeded no further. However, Fabre's lightweight, hollow wooden floats, which gave a measure of lift in the air as well as on the water, continued to be supplied in one form or another to hydroaeroplane manufacturers in Europe for several years to come. (A memorial to commemorate his first flight at Martigues was dedicated in 1967 by the Russian cosmonaut Yuri Gagarin—the first man to fly into outer space.)

At this time the U.S. Navy was showing signs of interest in the aeroplane as a marine scout. Eugene Ely, on November 14, 1910, attracted nationwide attention by taking off from a wooden platform erected on the

Eugene Ely's Curtiss biplane drew wide attention by taking off from the deck of the cruiser Birmingham during tests for the Navy at Hampton Bays, Virginia. (National Archives)

deck of the cruiser Birmingham at Hampton Roads, Virginia, and safely flying his standard Curtiss pusher to land two and a half miles away at Willoughby Spit. On January 18, 1911, as part of the show at an aerial meet, the twenty-five-year-old aviator reversed the performance, flying from land to the cruiser Pennsylvania, anchored in San Francisco Bay, and back again to his starting point. Nine months later Ely was dead, killed on October 9, 1911, while flying in an exhibition at Macon, Georgia.

But Curtiss did not believe that this kind of launching and landing typified the future of naval aviation. He was convinced that a hydroaeroplane—a stock Curtiss model fitted with a pontoon—would be more practical; it would also satisfy a preference in certain naval circles for a machine that could alight on the water alongside a battleship and be hoisted aboard, thus eliminating the "false deck" needed for landing on the ship itself.

From the date of his earliest experiments at Hammondsport, Curtiss had been considering the possibility of flying off the waters of Lake Keuka. In November 1908 he had vainly tried to turn the June Bug into a hydroaeroplane by equipping it with pontoons and renaming it the Loon. Late in 1910 he found what he thought was the ideal testing spot for such machines—at North Island, off San Diego, California. There the climate was favorable, sightseers were few, and both the Army and the Navy helped the Curtiss project by detailing officers for instruction in aviation.

A submarine officer, Lieutenant Thomas Gordon Ellyson, was the first to report for duty at North Island; he was soon joined by a Naval ensign—Charles Pousland—and by Lieutenants Paul W. Beck; John C. Walker, Jr.; and George E. M. Kelly, all of the U.S. Army. Ellyson, a red-haired, freckle-faced Virginian, blazed the trail for naval aviation: he was to receive Navy Aviator's Certificate No. 1 at Hammondsport on July 2, 1911.

Curtiss got his first hydroaeroplane model into the air on January 26, 1911. Beginning with his standard biplane, he had attached a short, wide pontoon where the two main wheels had been; a very small float in place of the front wheel; and inflated motorcycle tubes beneath the wing tips—a makeshift design that, however, proved unwieldy and was soon discarded. Next, a single large float was installed between the main wheels, greatly improving performance; and from then on, development was rapid. The Triad—so called because it was a universal vehicle, at home on sea, on land, and in the air—was a further development of the Curtiss hydroaeroplane. It had a pair of wheels that could be retracted when the aircraft was taking off from or alighting on the water, could be lowered for use as beaching gear, or could be extended and locked when the plane was to be flown from or to land. This model was ordered by the Navy, which gave it the designation "A-1" upon delivery at Hammondsport in July 1911. Several other Triads were commissioned by private pilots. For the Navy, its usefulness in scouting was obvious; for the tourist, pleasure voyages provided vistas over lake, river, bay, and sea.

On February 25, 1911, Curtiss made his first flight from water to land

A favorite with exhibition flyers, Glenn Curtiss's Triad was named for its ability to navigate equally well on land, on sea, and in the air. (National Archives)

and back to water. Starting from nearby Spanish Bight, he alighted on the beach close to the Coronado Hotel, a fashionable caravansary, and then returned to the calm stretch of water from which he had taken off. His description of this new branch of aviation was lyrical:

. . . flying a hydroaeroplane is something to arouse the jaded senses of the most blasé. It fascinates, exhilarates, vivifies. It is like a yacht with horizontal sails that support it on the breezes. To see it skim the water like a swooping gull and then rise into the air, circle and soar to great heights, and finally drop gracefully down upon the water again, furnishes a thrill and inspires a wonder that does not come with any other sport on earth.

No review of the early history of hydroaeroplanes would be complete without mention of an experiment by the Wrights. During the period when they were attempting, without response, to interest the U.S. Army in their machine, an international naval review was being planned at Hampton Roads, Virginia, as part of the Jamestown Exposition. The brothers were certain that the sight of their aeroplane flying over the assembled battleships would convince the Government of the utility of their invention. Accordingly, in 1907, they constructed a pair of cylindrical pontoons, equipped with hydrofoils—a very early use of these water-lift surfaces, invented by the Italian Forlanini in 1905. They mounted an engine and propellers, using their standard chain transmission, on a platform supported by the floats. The Wrights then experimented with this craft on the river near their home in Dayton. At about that time, however, the French government showed interest in their Flyer; and they set aside these aquatic investigations in order to prepare for their flight demonstrations in France. In 1915, when the Wright company at last produced a hydroaeroplane for the Navy, it was furnished with cylindrical floats and hydrofoils much like the gear devised in 1907.

From the start, the "hydro" not only furnished a thrill to the civilian populace but inspired wonder in the Navy Department, too. Captain Washington Irving Chambers, coordinator of the Navy's aeronautical activities, said: "The hydroaeroplane is the coming machine as far as the Navy is concerned; in fact it has already come." His assertion was lent substance by the Naval Appropriation Act for the fiscal year 1911–12, which granted $25,000 for aviation to the Bureau of Navigation. This enabled the department to purchase one Curtiss hydro of the Triad type, one Curtiss land model for training purposes, and one Wright biplane equipped with pontoons—the foundation of the U.S. Naval Air Force.

A year later, on January 10, 1912, an experimental Curtiss flying boat rose from San Diego harbor. "Flying boat" (sometimes called "aeroboat") was a term that differentiated such craft from a machine equipped with floats or pontoons; it more closely resembled a motorboat. In this case the machine featured two tractor propellers driven by clutch-and-chain transmission, with the engine forward in the hull. By July 1912, however, the typical Curtiss product in the flying-boat category had emerged. First tested on Lake Keuka, this two-seater, dual-control craft with a step in the hull had a single pusher propeller and engine mounted directly below the upper wing—a practice adopted by the designer from then on. Looking like a flying fish, the machine was powered by an 80-hp Curtiss engine and boasted a novel feature—a self-starter. Almost at once it obtained a great vogue with sportsmen who sought the sensation of pure speed obtained by skimming over waves and water rather than the joys afforded by ordinary aerial flight.

At a banquet of the Aero Club of America on January 27, 1912, Curtiss was awarded the Robert J. Collier trophy, presented for the first time (and thereafter annually) "for the greatest achievement in aviation in America, the value of which has been demonstrated by use during the preceding year." The trophy, representing the triumph of man over gravity and other forces of nature, was executed in the form of a bronze group by Ernest Wise Keyser of New York. It was the gift of Robert J. Collier, owner of *Collier's Magazine,* a genial and generous patron of flying who, almost alone in the United States, was prepared to encourage progress of the science by the donation of a substantial prize. Next year Curtiss again received the Collier trophy—this time for his development of the flying boat. He thus became the only person to take this award twice in a row.

The Curtiss hydro soon found its way abroad, where it did much to prove that in one respect at least America was abreast of the rest of the world in aviation. The Russian navy, for example, promptly indicated an interest in the Curtiss product, ordering several of the Model A-1 planes for its fleets in the Black Sea and the Baltic. The Japanese and Germans shortly followed the Russian lead. On May 11, 1912, W. B. Atwater of New York made three flights at Tokyo before Admiral Saito and a number of officials of the Japanese navy—the first ascent from the water ever witnessed in the Orient. Following the Atwater demonstration, the *Japan Advertiser* announced that an order had been placed for four Curtiss Triads, forming the nucleus of Japan's naval air arm. Germany, meanwhile, ordered two Curtiss planes: one for the Zeppelin works in Friedrichshafen and one for the Baltic fleet.

Glenn Martin was among the first to foresee the possibilities in over-water flight. On May 10, 1912, carrying an inflated tire tube as a life preserver, he piloted one of his planes fitted with a pontoon from Newport Bay, near Los Angeles, to Catalina Island—a distance of 38 miles—and back again. In this unprecedented oceangoing excursion he practiced the rudiments of instrument flying, with a compass strapped to one leg and a barograph to the other. Twenty-five years later to the day, in 1937, Martin was a passenger on a Pan-American China Clipper, the product of his own plant at Baltimore, in a commemorative trip over that first leg of the transpacific air trail he had blazed.

Martin's first experiments in overwater flying were with his own tractor biplane mounted on floats; the Clippers were lineal descendents of his ver-

Developed by an American yacht designer from the invention of a British army officer, the Burgess-Dunne hydroaeroplane was a tailless pusher with swept-back wings. (Author's collection)

sion of a hydroaeroplane, hailed in 1913 as a four-passenger "aeroyacht." This was the earliest promise of the success that was to follow for Martin aircraft of many types, the proof of which was to be the orders he received from the Government—running into billions of dollars—during World War II. Martin insisted in 1913 that the "aerial age" had begun; and this confident assertion was amply borne out by the triumphant progress he himself made over the next few decades.

The Wright Model B was also adaptable to water sport. In the hands of Frank Trenholm Coffyn, a Wright-trained pilot, one of these machines astonished New Yorkers in the winter of 1912 with its versatility. Resting on two pontoons made of a steel-aluminum alloy, it became known as the "flying toboggan and the aerial skate." An account in the New York *Herald* of February 7 told how Coffyn "demonstrated that his marine aeroplane can skate as well as fly and swim." It could "travel at will over rough, frozen masses inaccessible to an iceboat and glide without hesitation from water to ice or into the air at a turn of the rudder, all in a twenty-five-mile gale." Coffyn crowned his season by becoming the first to fly under the Manhattan and Brooklyn bridges.

In addition, W. Starling Burgess, a noted designer of yachts and speedboats, was turning his talents to the manufacture of pontoons and seaplane hulls. Burgess was already building, under license, Wright-type biplanes for the Army, and was presently to develop a tractor biplane as well. Now he began to experiment with a flying boat dubbed the Flying Fish, a type which he produced with an eye to purchase by the Navy. He also took on, under license from its inventor, the inherently stable tailless biplane of a British former army officer, John William Dunne. Dunne had given play to an active imagination by constructing a pusher biplane with extreme swept-back wings and short nacelle—almost like an arrowhead, minus its shaft, in appearance. Mounted on wheels, this unusual aeroplane met with an indifferent reception by the British War Office, and fared little better when produced under franchise in France by the Nieuport company. In the United States, however, it aroused considerable interest, especially when fitted with a single float by Burgess and used for over-water work.

Water, according to the hydroaeroplane enthusiasts, had numerous advantages over land as a base for flying operations. They theorized that water represented a nearly limitless natural aerodrome; that there were fewer air "pockets" over water than over land; and that water was "softer" to fall into than the hard and unyielding ground. A current nursery rhyme parody expressed the last consideration this way:

> Mother may I go out to fly?
> Oh yes my darling daughter.
> But do not go too near the sky
> And when you fall, hit the water.

Competing with individual hydroaeroplane developers such as Curtiss, Martin, Wright, and Burgess was the Benoist Aircraft Company of St. Louis, under the direction of the designer Tom W. Benoist. The Benoist firm, which transformed its tractor biplane into a hydro by substituting a single float for wheels, captured first-prize money at the Chicago meet of 1912. In the water events sponsored by the Aero Club of Illinois at Grant Park, chief Benoist pilot Antony Jannus led the Curtiss entrant of Beckwith Havens and the Curtiss copy of Glenn Martin by a substantial margin. At about the same time the company branched into the development of flying boats. Instrumental in this endeavor was Hugh Robinson—an up-and-coming flyer who had taken his license at the Curtiss school on June 25, 1911. Before joining Benoist, Robinson had made an ambitious attempt to stimulate interest in river flying, embarking on a four-day trip along the Mississippi from Minneapolis to New Orleans. Owing to lack of financial (or any other) support from cities en route, he was forced to abandon the project after covering a total of 314 miles.

The Benoist two-seater flying boat, powered by a 75-hp Roberts motor, was placed on the market at $4250. While in general appearance it did not differ materially from other such craft, it was claimed—not quite accurately—to be the only one in the world that had the motor set down in the hull, making it in effect a "motor boat with wings and an air propeller." Greater stability, and therefore greater safety, were said to result from this position of the motor; but the necessity of keeping the propeller high to escape the spray required that it be chain driven. As other designers found out, the power thus lost made the arrangement impractical.

Still another early American in the realm of overwater flying was

Three hydroaeroplanes
(biplanes)

Tailfirst Voisin

Maurice Farman

Curtiss (built by Paulhan)

Three hydroaeroplanes
(monoplanes)

Borel

Deperdussin
(built by Brouckère)

Blériot

Grover Cleveland Loening. In June 1910, Loening received a master's degree in aeronautics from Columbia University—the first of its kind awarded in the United States. A resident of New York, he began work with the Queen Aeroplane Company on Long Island, the firm founded by the wealthy Chicago stockbroker and aeroplane fancier Willis McCornick, who had come east to manufacture Blériot-type monoplanes. At the beginning of 1912 Loening concocted a flying boat out of a Blériot fuselage, a pair of discarded wings, and an old float. This was followed in June by a Queen "aeroboat," powered by a 35-hp Gnôme motor. A year later, in June 1913, Loening turned out an entirely new model, almost wholly made of metal. A monoplane with a 40-foot wing span, it was wrecked in a storm before it could be fully tested.

With this stretch of experience behind him, Loening went to Dayton as an engineer with the Wright company. Later he was to establish an aircraft manufacturing company bearing his own name. Eminently successful as a constructor and skilled as an aeronautical engineer, Loening foresaw the ascendancy of monoplane over biplane construction; but prejudice in favor of biplanes, for reasons associated with safety, was so great that he could find practically no supporters for the single-wing design that he had favored from the start.

The handful of Americans who preferred to be pilots of water-based rather than land-based aircraft had grown in 1913 to a point where the Aero Club of America began to issue "hydroaeroplane certificates" under rules of the Fédération Aéronautique Internationale. The first such certificate was issued to Alfred G. Sutro, and the next four to Navy aviators. By 1914 a total of sixteen hydroaeroplane licenses had been granted.

If Americans were quick to see the potentialities of water-based aircraft, Europeans were not slow to do so either. It was no great problem to adapt a land machine to water by the substitution of floats for wheels; and many of the leading firms and pilots in England and on the Continent went in for this variation of conventional plane design.

Thus, the *"hydro-avion,"* constructed by the Wright-licensed Société Astra, was essentially a tractor biplane mounted on two long pontoons—a machine which, like its counterpart on land, retained little but the wing-warping principle from the original Wright design. It was tried out first in Switzerland, on the calm waters of Lake Lucerne, by Maurice Herbster, a licensed pilot of the Farman school. Swiss lakes were natural "aerodromes" for such craft: a Blériot type was put on floats by Robert Grandjean at Lake Neuchâtel, while the Dufaux brothers zoomed over Lake Geneva in their homebuilt aquatic pusher biplane. E. Taddeoli, another Blériot pilot and the first Swiss to obtain an aviator's license, was badly injured in a fall at Lausanne in June 1911, but recovered to build a hydroaeroplane in 1912.

In England, Lake Windermere on the Scottish border provided a mirrorlike expanse for aquatic flying experiments in 1911 and 1912. The Short brothers biplane of N. Stanley-Adams, for instance, was so successfully—and noisily—tested that the Windermere Hydro-Aeroplane Protest Committee was formed against such an outlandish disturber of the peace.

Both the Henry and Maurice Farman planes could be converted by means of two long, narrow pontoons without steps. The Borel had a small float under the tail interconnected with the rudder, while the two sizable front floats were fitted with oarlocks so they could be rowed. The Caudron, with its distinctive short nacelle and double rudders; the Savary biplane; the Goupy, with its staggered wings; the Train monoplane, with the pilot under the wing and the motor at the leading edge; the standard Deperdussin; the Nieuport three-seater; and the R.E.P., mounted on one large, square central float, were some of those that took to the water—if not like ducks, then like hybrid fowl. The machine most resembling a water bird was the long-fuselaged 1912 Voisin *canard,* which utilized a new engineering concept in that its rotary Gnôme engine was mounted on the nose of the craft, in front of the propeller.

José Luis Sanchez-Besa, a pioneer Chilean aviator resident in Paris, produced two "hydro-biplanes"—the first with two and the second with three floats. A later version had a single boat body mounted on wheels—one of the earliest amphibians. Other amphibians to appear in 1912 were a two-seater naval *"motovoiturette"*—the small, robust Bedelia biplane, with chain-driven propeller in front, a four-cylinder Clerget motor in the metal hull, two wheels, and skids; the Marcel Besson *canard* monoplane, constructed entirely of steel tubing, with wing tips curved slightly upward; and a naval type, built by the firm of D'Artois, using either a stationary or a rotary motor of 50 or 100 hp, with four-bladed propeller, and with the power plant embedded at the center of gravity in the hull.

Among those who favored the flying boat rather than the hydro was the farsighted, many-sided Louis Paulhan—who founded a school at Juan-les-Pins, near Antibes, with an imported Curtiss Triad and who later built flying boats under a Curtiss license. The Donnet-Levêque was another French example of this type, on the order of the Curtiss in appearance: a single long hull with a step, pontoons at each extremity of the lower wing, and either a 50-hp or an 80-hp Gnôme motor. It was flown by Conneau of the French navy (the "André Beaumont" of the great races of 1911).

Some of these models—all biplanes—were displayed to a cosmopolitan public at the world's first hydroaeroplane meet, organized by the International Sporting Club of Monaco for the week of March 24 to 31, 1912. As in the case of the early land-based planes, the Riviera in spring was an ideal setting from the viewpoint of atmospheric conditions; and Monaco provided a permanent audience for the latest fashions in aviation. Against the theatrical backdrop of mountains falling sheerly to the sparkling blue sea, the newfangled water birds winged up and down the coast—even venturing out beyond the three-mile limit. Their pilots competed under the conditions of four categories: (*a*) takeoff in calm water, in the port, to circle marking buoys; (*b*) landing in calm water, in the port, coming in from the open sea; (*c*) takeoff in rough water; and (*d*) landing in rough water.

European aviators, like their machines, had become adapted to the new demands. Those present included Jules Fischer, a Belgian yachtsman and duck hunter, known in Italy for having flown his Gnôme-powered Henry Farman around one of the landmark towers of Turin; Eugène Renaux, with a Maurice Farman powered by the customary 70-hp Renault; Paulhan, with his Curtiss Triad; Hugh Robinson, representing the United States (the only non-Frenchman besides Fischer); René Caudron, and his light, Anzani-powered biplane; Jean Benoist (not to be confused with the American Tom W. Benoist), with a heavy Sanchez-Besa; Paul Rugère, flying a Voisin *canard;* and Maurice Colliex, with a second Voisin. Rated on a system of points, Fischer was the winner, with a score of 112.10. Renaux was second, with 100.80 points.

While the competition brought forth no new designs and resulted in no important records, it emphasized the comparative safety of the water-based vehicles. A few spills in a shower of spray—including a spectacular plunge into the ocean by Robinson—caused no loss of life or limb; dozens of paying passengers were carried without a single mishap.

The 1912 gathering stimulated an immediate response. As an incentive to seaplane construction and in encouragement of overwater flying, Jacques Schneider, leading patron of the sport, announced a valuable trophy—to be competed for annually like the Gordon Bennett cup for land-based planes. Inaugurated on April 16, 1913, the Schneider trophy

A Maurice Farman hydroaeroplane, entered by Renaux at the Monaco competition.

was initially won by Maurice Prévost in a Deperdussin fitted with floats and driven by a new 160-hp Gnôme motor. Special contests for this occasion included—for the first time anywhere—a rigorous 2½ nautical miles of surface navigation; for if overwater planes were ever to become practical, it was apparent that they must be at home on the sea as well as in the air.

The example set by Monaco at the winter's end was followed by two further events of importance in the summer of 1912. The first was a contest for speed, organized by the Automobile Club of France for August 24, 25, and 26 at St.-Malo, on the coast of Brittany: two races in St.-Malo Bay, and one race to the British isle of Jersey in the Channel and back, with an obligatory stop at Jersey. No fewer than ten different makes of machine were represented—for the most part flown by pilots whose experience had hitherto been over land.

For the first time, monoplanes were entered in an aquatic contest; and crowds converged on St.-Malo to see what the single-surface machines could do. Five of the eleven entries were monoplanes, all of French make: an R.E.P., piloted by Henri Molla, the new standard-bearer of the house; the Nieuport of Charles Weymann, a model which had given America the Gordon Bennett cup in 1911; an Astra-Train, latest vehicle of Emile Train, the unlucky pilot in the Paris–Madrid catastrophe; a Borel, flown by Marcel Chambenois—the first monoplane ever to be fitted with floats; and the Deperdussin of Guillaume Busson. To the delight of the monoplane's supporters, Molla's R.E.P. showed its tail to all other competitors in the opening day's events.

The biplanes, however, had to be reckoned with on a point system of scoring. These included the Maurice Farman of Eugène Renaux; a Donnet-Levêque flying boat, with Jean Conneau at the controls; two Paulhan-Curtiss Triads, piloted, respectively, by Frank Barra and a new flyer named Mesguich; the Sanchez-Besa of Jean Benoist; and a heavy Astra tractor, flown by René Labouret. It was a first-day victory on points for the Astra; able to carry two passengers with ease, this Wright-licensed machine was high scorer with the extra load.

Whether the meet could continue on the second day was a question—for a strong wind had sprung up and the sea had roughened considerably. Taxiing against the waves, some machines were damaged; others could not get off the water or threatened to capsize. Six planes finally took part in the day's trials. Those that managed to become airborne demonstrated that the double float, in catamaran style, afforded more stability than a single float. An exception, however, was the R.E.P.: with its broad single pontoon, it rode the swells without apparent difficulty—and for the second time Molla made the fastest record over the course. Again, however, a biplane was the winner on points, when the Sanchez-Besa of Benoist took off with three passengers and flew steadily into the freshening wind.

On the last day, angry whitecaps appeared, and the wind stiffened once more. Fortunately, for the race to Jersey and back it was permissible to

A Voisin hydroaeroplane being lifted aboard a French cruiser in Marseilles harbor, 1912. (National Archives)

land and take off from the sheltered harbor areas. Speed from point to point rather than number of passengers was what counted. Weymann won the race in his Nieuport, showing again that monoplanes were the swiftest craft on water as well as on land. Yet the high scorers of the meet were the biplanes: the Astra in first place, the Sanchez-Besa in second.

The next contest, which brought the season to a close, was organized by the Aéro Club de Belgique for September. Rather than court the risk of rough water, the meet was held at Tamise (or Tense), between Antwerp and Ghent, on the unruffled surface of the river Schelde. But the purpose of the encounter was unusual. Intrigued by the possibility of using water-based planes on the tortuous Congo River to knit together their colonial outposts in the Congo, Belgian officials planned to use the competition to determine what type of plane was best suited for this purpose. For the first time, an aquatic meet was held not for sport but to serve a pragmatic end.

Stringent tests were imposed, designed to reveal weight-carrying capacity, navigability, quickness of takeoff, and distance flown without stops; and since a commercial stake of national interest was involved, the entry list was a representative one. Nine biplanes and five monoplanes were submitted by eleven manufacturers—including for the first time a German Aviatik, the Autovia, with a 100-hp Aviatik motor, piloted by Bruno Büchner. Two homemade Belgian machines were entered, both converted from land to water: the 70-hp Gnôme-powered biplane of Alfred Lanser, one of the competitors in the recent Paris–Brussels race; and the Jero biplane, with a 50-hp Gnôme, of F. Verschaeve, another well-known Belgian pilot.

Considering the generally unfavorable weather, the trials took place with remarkable regularity and without noteworthy incident. Machines skimmed over the placid waters of the Schelde and took to the air without damage from waves; wind and rain were no serious problem; and the only accident recorded was one forced descent on hard land. An improved version of Védrines's victorious machine in the Paris–Madrid race—the Borel monoplane of Géo Chemet, with a Gnôme motor of 80 hp—proved its superiority over other types. Its high score once again demonstrated the advantages of the catamaran principle as against either a single-float machine or the flying boat with hull. Nevertheless a flying boat, the Donnet-Levêque of Conneau, took a special trophy donated by King Leopold. On the strength of this performance the firm extensively advertised its product as the perfect vehicle for overwater touring—featuring "very rapid instruction and . . . a minimum of danger."

Not until 1913 was sufficient interest aroused in the United States to warrant a contest for water craft. Under the auspices of *Aero & Hydro,* a Great Lakes "Reliability Cruise" was organized for the week of July 8—the course to follow the shoreline from Chicago to Detroit via the Straits of Mackinac. It was heralded as the biggest competitive aerial event for the year.

Most of the pilots who had taken up the practice of flying over water were on the entry list—a total of fifteen names. John B. R. Verplanck, an affluent sportsman from the Hudson River Valley, and his seasoned pilot, Beckwith Havens, entered a Curtiss flying boat with 90-hp Curtiss motor—as did Charles C. Witmer, Jack Vilas, G. M. Hecksher, and Navy Lieutenant John H. Towers. Antony Jannus, Hugh Robinson, and Tom Benoist entered Benoist flying boats, each with a Hall-Scott motor of 100 hp. Walter E. Johnson, who had worked as a mechanic for Glenn Curtiss, enlisted himself as the pilot of a Thomas brothers flying boat specially designed for the contest; with a 65-hp Kirkham motor, it was the first aircraft with an all-metal hull in the United States. Glenn Martin entered his tractor hydro with 90-hp Curtiss motor. Although labeled a "queer craft" by the Los Angeles *Examiner,* it had carried three passengers in California without trouble, and was headed for altitude records.

Others on the original list were Max Lillie (the first to receive an "expert aviator's certificate" from the Aero Club of America), piloting a Walco monoplane flying boat with 70-hp Sturtevant motor; DeLloyd Thompson, flying a Walco biplane model with 50-hp Gnôme; Roy Francis, with a Paterson tractor hydro powered by an 80-hp Hall-Scott; Weldon B. Cooke, with his Cooke flying boat fitted with 75-hp Roberts motor; and Frank Harriman, also with a flying boat and engine of his own make.

When the day of the race dawned—one of the stormiest in years on Lake Michigan—the list had appreciably shortened. Only five flyers actually managed a start from the Chicago lakefront either that morning or the next: Johnson, Jannus, Havens, Martin, and Francis—and only one, Havens, reached the first control point at Michigan City. Johnson, vainly fighting the weather, put in at Robertsdale, Indiana, only a short distance out of Chicago—while lifeboats searched for him until word came of his safety. From Michigan City to the control points at Muskegon (45 miles) and Pentwater (81 miles) beyond, the pilots had difficulty with rough water, balky engines, and broken propellers—the last a common complaint caused by damage from spray. Such obstacles slowed progress and kept public interest at a minimum. Holes were knocked in floats, and wind and high seas continued to harass the contestants—till, on the seventh day, only the team of Havens and Verplanck could be said to have made a creditable showing. Alone on July 14 they flew the distance of 138 miles between Pentwater and Charlevoix, in 2 hours 25 minutes at an average speed of 70 m/hr. On July 15 the race ended in recriminations—a fiasco as far as "reliability" was concerned. In view of the unexpectedly poor showing, the committee was reluctant to pay out prize money, while the prospect of flying without reward was not pleasing to the competitors. Verplanck and Havens finished in Detroit on July 18 and decided to prolong their Great Lakes excursion, giving exhibitions here and there; Martin announced that he, too, would exhibit independently; but Francis felt it was time to dismantle his machine and ship it home. All the others had given up. It was not a heartening experience for proponents of the hydroaeroplane in the United States—especially as the Schneider cup race at Monaco had just laid the foundation for record-breaking performances over water.

Americans could, however, take satisfaction in the fact that Glenn Curtiss had given the world the first flying boat—the development of which was one of the leading features of aviation in the last year before World War I.

CHAPTER NINE

Jesters of the Air

By the year 1913, the records for altitude and duration showed clearly that man could put himself aloft at will, and that he was gaining the confidence to fly higher and longer whenever he wanted. Now he began reaching for distance, as well as for speed and yet more speed. The great cities of Europe—and to a lesser extent those of the United States—served as goalposts for venturesome pilots seeking to set new marks for mileage or swiftness of flight. Tentative plans for commercial routes were already being considered. In some countries military authorities were becoming more than interested spectators; for the airplane began to show its potential as not only a vehicle for scouting but also a convenient and rapid means of transportation for high-ranking officers.

The worst of the winter was scarcely over when the public was treated to a series of spectacular long flights—the last in that peaceful era of sport and experiment before World War I. On February 25, Marcel Brindejonc des Moulinais, a twenty-year-old veteran of the Circuit of Anjou, started from the Villacoublay aerodrome, near Paris, on an aerial tour that took him first to London, with a stop at Calais; from London the next day back to Calais (during which he descended through the Channel fog to ask his way of a fishing boat); and, on the 28th, from Calais to Brussels and back to Paris—thereby completing, under difficult weather conditions, what the Paris *Herald* called "one of the most remarkable performances in the history of aviation." It had a remarkable sequel too: although flight plans, passports, and visas were still requirements of the future, Brindejonc was prosecuted under the Aerial Navigation Acts for flying into England without giving advance notice.

Brindejonc's name rolled off everyone's tongue—and the sturdy Morane-Saulnier monoplane with its 50-hp Gnôme motor, able to reach speeds of 110 to 115 km/hr, received lavish praise from the press as a "long distance" aeroplane. But that was only the beginning. On March 24 Brindejonc departed from Paris for Madrid, where he arrived on April 1, hampered by wind and rain all the way. Returning, he met with even more delay—leaving the Spanish capital on April 5 and reaching Paris only on the 16th—but he had chalked up some 3000 km of air travel. On June 10 he gave the most impressive demonstration yet of how nations could be brought closer by the aeroplane. Competing for the Pommeroy cup—to go to the flyer covering the greatest distance between sunrise and sunset—he flew 1450 km, from Paris to Warsaw, in 8 hours. On this astonishing journey, Brindejonc had breakfast in Paris, lunch in Berlin, and dinner in Warsaw—a prophetic indication of the modern international travel schedule. On the first stage of this flight his monoplane was whipped along by a storm at more than 2 miles a minute. When he tried to land at Berlin, the machine was hurled aloft by a furious gust after one wheel had touched the ground; and only by the exercise of great skill was he able to land safely on the second attempt. After he had gained Warsaw, Brindejonc pressed on into Russia. Alighting at St. Petersburg, he next crossed the Baltic Sea to the Swedish coast and Stockholm, then went on to Copenhagen and The Hague. From Holland he returned to Paris on July 2—having covered 5000 km in 22 days, with only seven landing spots en route. For that record Brindejonc received the Cross of the Legion of Honor and earned the fitting sobriquet "King of the European Capitals."

Paris was still marveling at the triumphs of Brindejonc when Léon Letort, on August 23, made the first nonstop flight from the French capital to Berlin. Letort also used a Morane-Saulnier monoplane, equipped with the new 80-hp Le Rhône engine—a seven-cylinder rotary that was making inroads on the popularity of the Gnôme. He covered the distance of more than 900 km in 8 hours, soundly beating the pioneer effort of Edmond Audemars (who a year previously, on August 18 and 19, 1912, had thrilled Europe by pushing his Blériot over the same route in two stages of approximately 450 km each). The new record dramatized twelve months of progress in long-distance flying—and also showed that the demands made on a machine for these new successes were becoming far more exacting than the Blériot's qualities could satisfy. Letort did not live long to enjoy his success; he was killed in France in 1913, crushed by the motor when his plane overturned.

Another contender for distance honors in 1913 was Eugène Gibert, whose legendary battle with an eagle in the Paris–Madrid epic of 1911 would always be associated with his name. Since that episode, during a stint in military service, Gilbert had suffered a bad crash; but in March 1912 he had recovered sufficiently to make one of the first city-to-city flights, over mountainous territory—from Clermont-Ferrand to Brioude and back. After touring France with a Sommer monoplane, he too had adopted the clean-cut Morane-Saulnier, with which he rounded off the year 1912 by breaking the speed record for all distances between 350 and 600 km. For variety, Gilbert began then to try for altitude with passengers, turning in fine performances with a Henry Farman biplane. Going back to his Morane monoplane, he established a nonstop record of 410 km in 3 hours 10 minutes between Paris and Lyons. And on April 24,

1913, in a bid for the Pommeroy cup, he left Paris at dawn; after flashing across almost the whole of France, he came down for fuel at the small Basque town of Vittoria, 850 km distant. But an exasperating delay in obtaining supplies prevented his flying farther than Medina del Campo in Spain—1020 km from Paris—before dusk and the rules of the competition called a halt to his attempt. It was nevertheless the first flight of more than 1000 km in a day—a mark broken less than two months later by Brindejonc des Moulinais in his historic dash to Warsaw.

Gilbert was determined to surpass his friend and rival. On August 2 he rose from Villacoublay, with the announced intention of flying to Cadiz on the southern coast of Spain. A storm over the mountains of Castille forced him down at Caceres, near the Portuguese border, after a flight of 1300 km—not quite enough to top the record. Undismayed, he resolved on another tactic. The rules of the Pommeroy contest provided that the prize could also be won by an aviator who covered 1000 km in a straight line in less than 5 hours. On the last day of the contest, October 31, gambling that he would not be forced down, Gilbert set out from Paris in a racing Deperdussin with clipped wings, which he had used in the 1913 Gordon Bennett race at Rheims and which required more than a kilometer of runway for landing. This time he took off in a different direction—and reached Putnitz, in German Pomerania (now Poland), in 5 hours 23 minutes for a total of 1200 km; but having changed course over the Baltic, he was disqualified for the straight-line classification. Gilbert's disappointment was offset by having placed third for France in the Gordon Bennett race that year—and later when he put in a bid for the Michelin cup of 1914 with a 3000-km circuit of France, through fog and rain, in a total time of 39 hours 35 minutes, including stops.

There remained one great challenge for Europe's birdmen: the crossing of the Mediterranean. To span the 474 miles of sea that separated France from the continent of Africa was a formidable undertaking, compared with Blériot's cross-Channel hop only four years previously. But the feat would bring added glory to France; and Roland Garros, looking around for new worlds to conquer, decided to have a try. It would be a chance to outstrip even the success with which he had wound up the year 1912. In a duel with Georges Legagneux for the world altitude record, he had reached 5610 meters in the mild air of Tunis on December 11. And a week later, on December 18, he had flown from that city to Trapani, Sicily—an aerial excursion of 285 km that linked two different continents of the world for the first time. From Sicily, Garros had flown on to Rome for a total distance of 1200 km, half of it over water.

Like many other flyers, Garros had now given up the light and relatively delicate Blériot in favor of the heavier and stronger Morane-Saulnier, with its 60- or 80-hp Gnôme motor. One of these machines, equipped with floats, had given him second place—after Maurice Prévost and his Deperdussin—in the first Schneider cup race at Monaco the previous spring; and with another he had taken part in the second annual meet at Vienna.

The crossing of the Mediterranean by aeroplane had been attempted before—but not successfully. Lieutenant E. Bague of the French army, becoming captivated by the novelty of flying over water, had asked for an assignment that would help to further his ambitions. Refused by the authorities, whose caution was to prove wholly justified, Bague turned in his resignation and proceeded to qualify for a private pilot's license at Pau on December 20, 1910. The Mediterranean flight became for him an obsession. Finally, at 7:30 on the morning of March 5, 1911, Bague took off from Nice in his Blériot and headed for Corsica—the nearest stepping-stone to Africa. It was not long before he was lost in a fog. But miraculously, after 210 km of flight, he found the tiny islet of Trigona—a mere speck in the sea to the east of Corsica—where he landed safely. Despite the hazards of the flight thus revealed, Bague persisted in his plan. On June 8 he started out again, taking with him several carrier pigeons to communicate with land in case of a forced descent. Nothing again was ever heard of him or his pigeons. After eight days of intensive search by torpedo boats among the small islands off Corsica and their surrounding waters, Bague and his machine were given up for good.

Two other examples of the risk in bridging even a short expanse of water were at hand. On December 20, 1911, Cecil Grace, nephew of the sometime mayor of New York, took off from Calais to cross the English Channel. He wore a cork life belt, and his Short biplane carried a fuel supply good for five hours; yet the only clue to his fate was a leather helmet, with goggles attached, which subsequently came ashore on the Belgian coast. And in April 1912, Denys Corbett Wilson, an Irishman who had learned to pilot a Blériot at Pau, set out to fly from London to Dublin. He was last seen starting across the Irish Channel at Holyhead; after that all trace of him and his machine was lost.

Although the science of aviation had by now made conspicuous advances, the odds confronting Garros were anything but reassuring. No such extended journey over lonely open water had ever been made. Even with the safe havens of Corsica and Sardinia attained and passed, there would still remain five or six hours before the Tunisian coast would be sighted. The Morane-Saulnier, a land-based machine with rigid rather than swiveling undercarriage, could carry gasoline for no more than eight hours of normal flight; and in case of a forced descent it would certainly sink. There were to be no naval units on the route. (Garros felt they would only handicap his freedom in choosing the right moment for departure.) A head wind blowing up from the south, or a slight error in navigation that extended his trip by minutes, could spell disaster.

But Garros—calm, confident, and methodical in his preparations, leaving nothing to luck—was undeterred by worried friends who sought to dissuade him from what they considered a blind leap in the dark. Shortly after daybreak on September 23, 1913—at 5:27, to be exact—he took off from the beach at Fréjus, near the scenic Riviera town of St.-Raphaël. He did not wear a life preserver but carried a rubber tire inner tube instead. After 7 hours 53 minutes of steadfast flying—fraught with anxiety after damage to a cylinder head in flight—Garros saw the low-lying coastline of Tunisia dead ahead. Mingled with his relief was satisfaction that he had pressed on, despite his limping motor, and not yielded to the temptation to make an emergency stop at Sardinia, halfway. When he landed in the glaring sun at 1:15 P.M. on the parade ground of the French naval base at Bizerta, Garros had barely 5 liters of fuel left—enough for only a few more minutes of flying. No one had expected his arrival, and no one was on hand to greet him. But at the officers' mess that evening he was more a hero than a guest. Upon his return to France, Garros was tendered the homage and admiration of the world for an act unsurpassed in originality and nerve until the heroics of wartime flying.

The Morane-Saulnier monoplane used by Garros for the crossing was selected also by Marc Pourpe, a specialist in pioneer air explorations of the French colonial possessions. After flights in Europe, India, and Indochina, Pourpe conceived a daring project: to fly from Cairo to Khartoum and back—a round-trip distance over the desert of some 4500 km. The exploit was planned with the advice and encouragement of Britain's governor-general of the Sudan, Lord Kitchener. The trip out was successfully completed in six stages between January 4 and 12, 1914, in 16 hours 18 minutes of flying time. The return journey was accomplished more leisurely.

When World War I erupted, Pourpe gave up the most ambitious plan of his life—to fly from Paris to Saigon, French Indochina, and thus link France with the outermost limits of its empire. Instead he joined the squadron in which his friends Garros and Gilbert were serving at the front. He was forced to leave behind his constant companion in the open cockpit of his monoplane: an air-minded brindle pup.

Progress in overwater flying continued up to the eve of war. The North Sea—traditionally a wild and stormy stretch, traversed by few ships—offered one of the sternest possible tests. Tryggve Gran (born in Bergen, Norway, in 1889), an associate of the Antarctic explorer Robert Scott, resolved to conquer the challenging expanse of water. In 1914, Gran signed a contract with Blériot for one of the latter's used machines, equipped with an 80-hp Gnôme motor—to be delivered, as soon as it had been reconditioned, for a price of 13,000 francs. His preparations were quiet and efficient. On July 30, taking off from Cruden Bay, on the Scottish coast, without fanfare, Gran made the eastward crossing to Jaeren, near his birthplace in Norway, in 4 hours 10 minutes. During this time he was completely lost to the world; but his navigation was perfect, and the flight was completed without incident. A stone marker indicates the spot where Gran landed. And the historic monoplane Nordsjøen ("North Sea"), which realized the dangerous crossing, hangs proudly today in the Norwegian Technical Museum in Oslo.

In Germany the counterpart of Issy-les-Moulineaux was busy Johannisthal, a rectilinear flying area to the southeast of Berlin. Easily reached by train or road, the field was bordered on one side by pine woods but otherwise surrounded by open countryside, which was reassuring in case of engine failure. Inaugurated in 1909 with a "Flugwoche" ("Flying Week"), from September 26 to October 3, that promised prize money amounting to the substantial total of 150,000 marks, it experienced consistently successful seasons, as disclosed in the following table:

Dates	Event	Prizes (marks)
	1910	
May 10–16	Flugwoche	57,300
August 7–13	Flugwoche	28,600
October 9–16	Flugwoche	74,000
	1911	
June 4–11	Flugwoche	30,800
June 12–July 10	Circuit of Germany	(special)
Sept. 24–Oct. 1	Flugwoche	40,000
	1912	
May 24–31	Flugwoche	40,000
June 9	Start of Berlin–Vienna race	77,000
Aug. 31–Sept. 1	Circuit of Berlin	60,000
Sept. 29–Oct. 6	Flugwoche	42,000

On the calendar for 1913 were a spring Flugwoche, to run from May 25 to June 1; a Circuit of Berlin, on August 30 and 31; and an autumn Flugwoche, from September 28 to October 5. By the opening of the spring Flugwoche, under the patronage of Prince Friedrich Leopold of Prussia, Johannisthal had expanded into a bustling, businesslike establishment that promoted flying not only as a professional and amateur occupation but also as a rewarding spectator sport. No fewer than eleven entrances gave access to different parts of the field; the facilities included three grandstands, a glass-enclosed restaurant, an extensive promenade, quarters for the Kaiserliche Aero-Club and the Berlin-Aldershof flying school of the Deutsche Luftflottenvereins, an air-travel bureau, a telephone exchange, post and telegraph offices, a meteorological station, automobile parking space, and administrative offices. To help accommodate the weekend and holiday throngs—especially large in fair weather—the covered main grandstand alone had seats for 2344 persons and boxes holding 1064.

Around this complex a group of prominent constructors and flying schools had set up shop. In addition to the usual factory buildings and sheds housing the flying machines, two large hangars for dirigibles dominated the skyline. Among the aircraft firms represented (all of which provided flight instruction) was the E. Rumpler Luftfahrzeugbau, makers of the stunningly successful Taube—which in 1911 had won the first German reliability contest as well as the Munich–Berlin Kathreiner prize, worth about $12,500. Its pilots included Frank V. Eckelmann, Joseph Suvelack, and the pipe-smoking Gino Linnekogel, a specialist in duration flights, who flew the Taube with passenger for a record-breaking 4 hours 34 minutes on December 8, 1911. Other firms were the Fokker Aeroplanbau; Emil Jeannin's Flugzeugbau—personally supervised by that industrious builder of monoplanes; the Luft-Verkehrs-Gesellschaft, producers of the excellent L.V.G. tractor biplane; the Harlan Werke, constructors of the low-wing, Argus-powered Harlan monoplane—whose engineer-pilot, Karl Grulich, had set a world duration record of 1 hour 35 minutes for pilot with three passengers on January 25, 1912; the Aero Flugzeugbau, run by Hermann Reichelt; the Flugschule of Frau Melli Beese-Boutard, an aviatrix-builder who obtained her pilot's license in September 1911; and the Flugmaschine Wright Company, which also numbered a woman among its pilots—the Countess Eugenie Schakowsky, a certified flyer since July 1912. Best known of all was the Albatros-Werke, maker of a Mercedes-powered biplane that was highly responsive to its controls but relatively undisturbed by wind gusts. Its chief exponent was the young engineer and racing car enthusiast Hellmuth Hirth, first-prize winner in the 1912 race from Berlin to Vienna and a regular participant in numerous other competitions.

Still other names on the hangars were those of Bruno Scholz, J. Merx, M. Conrad, the Deutsche Flugwerft of Dr. Fritz Huth, and such private experimenters as Dr. Geest and Paul Westphal—these last two employing Oskar Roempler as test pilot. Separate flying schools were maintained by the Allgemeine Fluggesellschaft, Bruno Hanuschke, Sport-Flieger, and Paul Schwandt.

With typical German thoroughness, the aerodrome authorities analyzed flight statistics for 1911 and 1912 under the following headings: date, number of days flown, number of flyers, numbers of flights, duration of flights, number of flyers receiving licenses, number of flyers making cross-country flights, and number of planes damaged in accidents. By the middle of 1913 the roster of craft registered at Johannisthal bore the names of 125 different pilots or owners; activity was increasing rapidly when war put a stop to Flugwoche and plane-watching alike.

Countess Eugenie Schakawsky, certified pilot of the Flugmaschine Wright Company.

Bruno Hanuschke, German pilot, constructor, and flying-school operator, in the cockpit of his plane. (Paul Nortz collection)

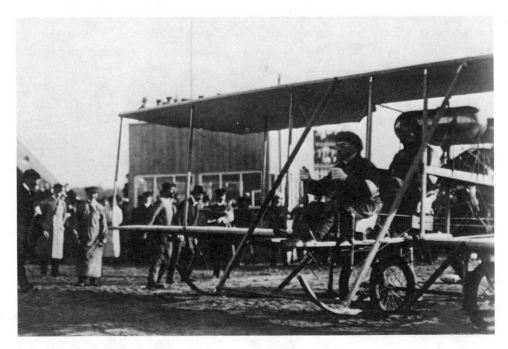

German pilot Robert Thelen preparing to take off in an Albatros biplane at Johannisthal.

Among the conspicuous characters associated with the early days of Johannisthal was Anthony (Tony) Herman Gerard Fokker, the son of a former Dutch coffee planter in Java, who had sent him to technical school in Germany, where he showed marked aptitude for mechanics. Inspired, like so many others, after seeing a Wright demonstration, Fokker resolved to teach himself to fly. In 1910 he tried out a primitive homemade monoplane, using the Zeppelin hangar at Baden-Baden as a base. Early in 1911 he moved to a field at Mainz, near Frankfort, where the Wright pilot Robert Thelen was flying. On May 16 Fokker earned his German pilot's license.

The Fokker monoplane of that era was a light, low-wing affair of wood, cotton cloth, and piano wire, which took off at a sedate 20 or 25 m/hr and attained a top speed of 40 m/hr. It was inherently stable and remarkably safe, using no ailerons or wing warping, with V-shaped, swept-back wings and a high center of gravity.

To move to the center of action at Johannisthal was a logical next step; and by the end of 1911—at the age of twenty-two—Tony Fokker had joined the crowd of student and exhibition flyers, designers, experimenters, and pilots of more or less proficiency who lived, often precariously, solely for flying. A likable fellow with a boyish smile that revealed rather large front teeth, Fokker's youthful appearance and abysmal inexperience were at first a joke to more seasoned birdmen. But their laughter changed to respect when the young Dutch flyer earned a disproportionately large share of the Sunday gate receipts—which were divided among the per-

An Anthony Fokker monoplane making a bank over the Johannisthal field, the German counterpart of Issy-les-Moulineaux.

Geest's Sea Gull, named for its flexed wings, in flight over Johannisthal.

formers in accordance with the number of hours they spent in the air. Up at five every morning to make the most of the hour or two of calm after sunrise, working hard in the hangar during the day, he won the applause of the animated group that assembled every evening, exuding bravado, at "Papa" Senftleben's conveniently nearby café.

The second Fokker machine was a two-seater trainer type powered by a 100-hp Argus motor. A skeletonized monoplane, its wings were supported by eight wires (four above and four below) on each side of an open fuselage. The new model responded so well to the controls that Fokker tried to interest his government in Holland in buying it; but the Dutch military were already sold on the Henry Farman. Nor were the English or the Italians willing to take a chance. It was the Germans from whom he received encouragement: the army ordered two planes, at 10,000 marks apiece.

Fokker went to St. Petersburg for the first Russian military trials in the summer of 1912, competing with the crack Russian stunt flyer Wssewolod Abramovitch, leading pilot of the Wright school at Johannisthal. (Abramovitch, whose hair-raising banks and turns with the Wright sent gasps of admiration through the crowds, came to a predictable end early in 1913.) Although Fokker did his best at the trials, the first three places were a clean sweep for the Russians.

It was hard going for Fokker financially—and without aid from his father he might have been forced to quit. Yet recognition was close at hand. In 1913, Adolphe Pégoud jolted Johannisthal with his unheard-of loop-the-loop; and Fokker was the first in Germany to copy the French master. It was clear that a new phase had been reached in the art of flying: the concept of an inherently stable plane had been outmoded by the

Dr. Fritz Huth's all-metal Taube-type monoplane at Johannisthal.

rapid increase in power, which gave pilots the control—and therefore the ability to maneuver—that they had previously lacked. Fokker abandoned his slow, safe, and stable plane in favor of a more conventional design— first warping the wings, then installing ailerons for lateral stability. A rectangular fuselage of welded steel tubing gave strength; the wings were of wood, covered with fabric, and braced with cables instead of wires.

With this model the Dutchman went on looping and stunting tours throughout Germany. "Now you are famous," wrote his father, "now is the time to stop. The only thing you can do next is to break your neck." Narrow escapes, so typical of the times, were frequent and dramatic. Once, spiraling to earth with gasoline from a burst tank streaming down the exhaust pipe, Fokker's plane exploded into flame seconds after landing. On another occasion he lost a wing in mid-air when his propeller became fouled in a broken guy wire; the crash killed his German army passenger. After that he used cables for his main bracing instead of the less reliable piano wire. But Fokker never broke his neck; he lived to see his name a synonym for one of the greatest of the World War I planes, and later a great commercial transport.

As Johannisthal became more congested—with sixty or seventy regular pilots, as well as army officers, learning to fly—it became more dangerous. In January 1913, there were 824 flights; in February, 2373; in March (reflecting inclement weather), 1564; and in April, an unprecedented 3224. Twenty-seven serious accidents were counted in the first three months of that year. Worse yet, on May 14, a disastrous collision took the lives of Captain August Jucker of the Luft-Verkehrs-Gesellschaft and his passenger, a student flyer named Dietrich. The L.V.G. biplane of Jucker was rammed amidships by Rubin Wecsler, one of the two principal pilots of the Harlan monoplane, who miraculously escaped death in the crash.

In his recollections of that period, Fokker wrote:

Relics of the crashes used to adorn the walls of Tolinsky's café, the mechanics' gathering place near Senftleben's. Indeed, much interest centered on early aviation depended on the lively prospect of a Roman holiday ending air meets. Attendance always picked up after a particularly messy crackup. We early flyers knew this, accepted it, even capitalized on it; though I think we were something more than gladiators.

Fokker decided to move to Schwerin, in northern Germany, where he rented a field, established a school, and built a factory that was soon to achieve renown. The German army, by now definitely interested, gave him a series of orders that kept him fully occupied and laid the foundation for the "Flying Dutchman's" association with the German side in World War I. Ironically, a question was put in England's Parliament after the war as to why this stupendously effective creation of a Hollander had not been acquired first by the Allies. The answer was, of course, that it had been offered but had received no takers. It was Fokker's passionate devotion to aviation that caused him to throw in his lot with the only government to assist him in his experiments; and no rancor is borne him on that account in his native country today.

Whatever the other records of the day, they were eclipsed by the climactic event of 1913: Maurice Prévost's performance in hurtling around the Bétheny aerodrome, near Rheims, at more than 120 m/hr, in the fifth race for the Gordon Bennett cup. It was an extraordinary demonstration of speed—of how fast, literally, aviation had advanced since 1909, when Curtiss took first honors for America at less than 47 m/hr. It was also the last great contest open to the nations of the world before war caused the abandonment of further competitions.

The starters in the Coupe Internationale d'Aviation on September 29 numbered four—three French and one Belgian. In addition to Prévost (who had been second in the 1912 contest at Chicago), two other noted aviators were on hand to defend French possession of the trophy: Emile Védrines, brother of the 1912 winner; and Eugène Gilbert, persistent pursuer of long-distance records. Henri Crombez, the fair-haired, twenty-year-old holder of Belgium's altitude record and inaugurator of his country's first aerial post during the recent Exposition of Ghent, was the only other competitor. All but one flew the latest version of the Deperdussin *monocoque*: the motor a powerful 160-hp Gnôme or Le Rhône; a Chauvière propeller; the wings clipped to the limit of safety; and the chassis carrying rakish-looking disk wheels. (A retractable undercarriage had not yet been invented.) The exception was Védrines, whose swift monoplane, also with cut-down wings, had been built by Alfred Ponnier,

Henri Crombez, the Belgian representative in the 1913 Gordon Bennett cup race, shown in a Deperdussin monocoque. (Belgian Institute, Brussels)

former director of the Hanriot establishment at Rheims and now a constructor in his own right. Védrines, chief pilot of the Copin-Revillard school at Juvisy, had joined Ponnier early in 1913.

Again until the last moment, no American effort was made to submit an entry. Then Norman Prince—soon to fall in combat flying for World War I's famed Lafayette Escadrille—offered to underwrite the purchase of a plane to represent the United States. But again it was too late for the plan to materialize.

Perfect conditions favored the race—little wind, a cloudless sky, a sun shining with almost midsummer warmth. Nostalgic memories of 1909 were evoked as thousands of spectators from Rheims, Paris, and more distant points poured in on Bétheny, clogging the dusty roads with automobiles and every other available conveyance. What they were to see, however, was like something from another world, compared with the puny attempts of four years before.

By the choice of the draw, the first to start was Crombez. Only the day before, he had found a Gnôme of 160 hp to replace the 140-hp model in his Deperdussin; and this he hoped—even if it was not fully tuned up— would give him the extra speed needed to beat the French. Crombez had made his first flight as a boy passenger with Louis Paulhan in 1909; his enthusiasm for aviation had been supported by an indulgent father, who had bought him not only the *monocoque* racer but two ill-fated Sommer

Henri Crombez seated in his Deperdussin monoplane. (Belgian Institute, Brussels)

monoplanes before that. The solicitude of the elder Crombez for his son's welfare led him to stay up all night with the mechanics while they changed the engine.

A few minutes before 10:00 A.M., a squad of gendarmes galloped across the field to chase away photographers too near the line of flight; and precisely on the hour the signal for departure was given. Crombez's mechanic turned the propeller amid cheers from the crowd, and, in the words of one reporter, the monoplane "darted forward like a bird released from its cage."

Circling the first pylon, as prescribed by the rules, the youthful Belgian then crossed the official starting line on his first dash around the 10-km circuit. He had been warned of another rule: he would have to tick off each lap for himself—though his mechanic would be allowed to give him a sign when the twentieth and last round had been completed. The official time for the first circuit, announced on placards immediately after the aviator had passed the grandstand, was 3 minutes 29 seconds. Considering the speeds expected from the others, it was obvious that Crombez would have to do better to win; and interest shifted to the next starter.

At 11:15, Marcel Prévost sped off in his Gnôme-equipped racer, its wings so reduced in size as to be almost invisible on the backstretch. "No words can describe adequately the impression created by the marvelous velocity of the French machine," reported a special cable to the New York *Herald* under a Bétheny dateline that day. Prévost flashed around the course at low altitude with clockwork regularity—his second, third, fourth, and fifth rounds completed in exactly 2 minutes 50 seconds each. At the end of the twentieth lap, a "deafening roar" from the spectators nearly drowned out the noise of the motor: he had covered the 200 km in the astounding time of 59 minutes 45⅗ seconds—or at the rate of 204 km/hr. No one had previously flown that distance in less than an hour. Prévost received a wild ovation when, after making another round to slow down, he brought his tiny craft safely to earth.

Eugène Gilbert, the next to ascend, was unable to equal the Prévost record. His Deperdussin with its 160-hp Le Rhône required 62 minutes 55⅖ seconds to complete the course—slightly longer than his teammate. Each minute and each second were now of decisive importance. Védrines then followed in the Ponnier monoplane. This machine had shown great promise in the elimination trials, and its pilot had set his heart on victory; but the 160-hp Gnôme-powered craft could not quite nose out the Deperdussin. Speeding regularly past the grandstand every 3 minutes, the Ponnier made the twenty laps in 60 minutes 51⅖ seconds, at the rate of 198 km/hr, to gain second place. Less than a minute and a half separated Védrines from the winner—and only a shade of difference could be found in the times for Gilbert, in third place, and for Crombez, in fourth.

Védrines was cheered over and over again for his effort, and shared with Prévost a triumphal welcome when both arrived at the buffet to celebrate. In the best champagne of the region, the victory of French flyers, French machines, and French motors was toasted by a throng of jostling admirers. Barely six months later, on April 1, 1914, Emile Védrines was to crash to his death on the same spot where he had come so close to winning the famous contest.

"Greatest Air Race Ever Known," proclaimed the New York *Herald*. ". . . the winner's speed being the fastest ever made by man in this form of locomotion. . . ." Declared Prévost, as quoted in the dispatch: "I am delighted at my success . . . the machine was in perfect flying order. The motor never gave me a second's anxiety." He added the prediction that with more powerful motors, "more wonderful speed could be obtained." In England the *Daily Mail* burbled that the "performance eclipses anything of the kind yet achieved," and that if the "speed could be maintained the Atlantic itself could be crossed in its narrowest part in a fraction over fifteen hours." The *Daily Telegraph* offered a less optimistic glimpse into the future: "It is doubtful whether, under next year's regulations . . . such velocities will be equaled, or for several years to come."

The failure of America to participate in the 1913 race again pointed up the slow pace of progress in the country whose representative had defeated the best that France could offer in the first Gordon Bennett contest. In a statement to the press Alan R. Hawley, president of the Aero Club of America, ruefully admitted the facts: "We could not send an American monoplane or biplane over, because none of our machines are half speedy enough."

Claude Grahame-White was openly bitter about the absence of Anglo-Saxon contestants:

Many of us would have given an eyetooth to have been in the race. I'd have been there like a shot if a machine had been provided for me. . . . But the supporters of aviation in England did not rise to the occasion, and so here I am, sitting at my desk and reading about the precious victory for France.

England should have been represented and America, too, and I daresay American aviators were as anxious to compete as we were, but those upon whose backing we relied were dead asleep.

Sportsmen in England and America should have prepared for suitable representation months ago. . . . It would seem that if aviators can be had to pilot machines some organization like the Royal Aero Club might show a little interest, but the entire burden of putting out money is shifted to the shoulders of the aviator himself. I am not speaking of France, for there the real sportsmen's spirit is shown.

While the marvelous machine to which he had given his name had again captured the headlines, financial and legal problems were rapidly closing in on Armand Deperdussin. It was the last time that he was to see his monoplane prevail; for the firm shortly went bankrupt. Out of the defunct company, and under the presidency of Louis Blériot, grew the Société Provisoire des Aéroplanes Deperdussin. This name was soon changed to Société Pour l'Aviation et ses Dérivés and abbreviated to SPAD—initials to be made famous by the World War I fighter plane they designated. Thus, from the ashes of a record-breaking pioneer racer rose, like the phoenix, a machine that was to help save France from disaster. Armand Deperdussin himself, in the most dire straits, was shortly to die by his own hand in a Paris hotel.

The next race, scheduled for September 1914, was canceled by the outbreak of war. For that event the United States and Great Britain—as well as France, the defending country—had each entered three planes; Germany and Italy had entered one apiece. The competition was not resumed until September 28, 1920, when a third victory for France gave that country permanent possession of the trophy.

A revolutionary new dimension was added to aviation in 1913 by a maneuver hitherto undreamed of: flying upside down, or "looping the loop." Such defiance of the elementary rules of safety had never before been thought possible, much less attempted. In an era when much store was set in an aeroplane's "inherent stability," topsy-turvy antics in the air seemed nothing less than suicidal.

It fell to another Frenchman to furnish the public with proof that an aeroplane could be controlled in inverted flight, turn a complete somer-

Autographed photograph of Tom Sopwith, British flyer and manufacturer, in the cockpit of his aeroplane.

sault, and perform other unorthodox feats without death or injury to the occupant. For until it was shown that it was not only possible but safe to fly in any position, Levavasseur's assertion that man could do better than the birds would remain unrealized.

In his book *Les Conquérants du Ciel*, Roger Sauvage notes that just as Saint Hubert is the patron saint of the hunt, so was Adolphe Pégoud regarded as the patron saint of fighter pilots in World War I. A short, round-faced man with an upcurled moustache and laughing blue eyes, Pégoud was born in 1889 in the Val d'Isère, near Mont Blanc. He made his initial flight in an R.E.P. as a mechanic during his military service. Shortly after joining the Blériot staff early in 1913, he made news by deliberately abandoning one of the company's machines—a rickety, Anzani-powered craft condemned to the scrap heap—at its ceiling of 250 meters and floating to earth in a parachute designed by the inventor Bonnet, while the old Blériot crashed to destruction. It was not the first time that a man had bailed out from a flying machine: U.S. Army Captain Albert Berry had jumped from the Benoist biplane of Tony Jannus near St. Louis on March 1, 1912, at an altitude of 1500 feet; and "Tiny" Broadwick had made history by dropping from Glenn Martin's machine near Los Angeles on June 21, 1912—the first woman to jump from an aeroplane. But Pégoud was the first pilot to leave a plane in flight, and his stunt brought him instant fame; it also paved the way for the wartime practice of parachuting from a machine shot down in combat.

Next, on September 2, 1913, at the Juvisy aerodrome, Pégoud coolly turned his machine upside down and flew before a selected—and astounded—group of friends and officials, who could scarcely believe what they saw. Then, on September 21, using a reinforced Blériot with a 50-hp Gnôme motor, Pégoud—by now known as "the Foolhardy"—electrified

Magician Harry Houdini in a Voisin-type biplane.

the world by looping the loop. The pattern he set was quickly repeated by others: P. Hanouille, another Blériot pilot, was the second; Maurice Chevillard, in a Henry Farman biplane with an 80-hp Gnôme, copied the maneuver in a pouring rain; P. Chanteloup, in a Caudron, was the first to do so with a passenger. By the spring of 1914, more than fifty aviators in various parts of the world—including twenty-eight Frenchmen, eleven Englishmen, and the American Lincoln Beachey—had achieved what could no longer be considered a miracle. Pégoud embarked on a highly publicized tour to Hendon in England, to Hannover in Germany, and to Belgium, Italy, Rumania, Russia, and Holland; he was about to leave for the United States when war changed his civilian career into a military one, in the service of France.

While Pégoud gathered world renown for the risks he ran, the evidence shows that an audacious young Russian officer, Lieutenant Peter Nicholaivich Nesteroff, looped his Nieuport a few days before the Frenchman—at the aerodrome of Kiev on September 9, 1913. Instead of praise, however, Nesteroff received the censure of his superiors for executing the "death loop." Uncomprehending of the significance of acrobatics to the future of aviation, the rigid-minded officers sternly charged the pilot with endangering government property and placed him under house arrest for a month. Later promoted to captain, Nesteroff met a hero's end in one of the first dogfights of World War I. The name of the Russian town above which the duel took place was subsequently changed to Nesteroff in his honor.

The contributions of Nesteroff and the much more celebrated Pégoud were to show the way to achievement of complete maneuverability in the air, to development of the tactics necessary for aerial combat, and to techniques of strengthening machines so that they would not crack up under undue stress. Like the races and competitions, which were in reality public testing grounds for the planes and pilots of competing manufacturers, aerial acrobatics tested and advertised a plane's general capabilities. They also furnished a way of life to daredevils attracted to the resulting prize money, and helped to keep the public air-minded.

Stunt flying was in fact the only sector of aerial activity that could still catch and hold the jaded Americans' attention. The brilliant performances of French and other foreign flyers during the summer of 1913, the steady improvement in their planes as evidenced by the Deperdussins at Rheims, were in sharp contrast to the parlous state of aviation in the United States. Still without the financial incentive to develop new machines, and with little or no Congressional interest in the aeroplane's military potential, America continued to lag far behind Europe in the air. Its contributions to the list of new marks in duration, altitude, speed, and distance were few and insignificant in comparison with the series of record-smashing flights elsewhere; and in the absence of exciting contests, public support waned. Flying as a sport was almost a dead duck.

That it was kept alive at all was due chiefly to such spine-tingling exhibitions as those of the legendary Lincoln Beachey—the greatest stunt pilot in the United States, if not the world. Born in San Francisco in 1887, Beachey began his career in the air as an operator of small dirigibles with Captain Thomas Baldwin and Roy Knabenshue. He graduated to exhibition flying with a Curtiss pusher, after taking a license in 1911; then, acquiring fame as an exponent of precision stunts, he crowned his reputation by looping the loop over San Diego Bay on November 24, 1913.

Short, clean shaven, a jaunty young man who disdained special flying clothes and flew usually in a business suit with cap turned backward, Beachey seemed a lonely and fearsomely exposed figure in the little Curtiss. Hands clasping the steering wheel, feet clamped behind the front wheel of his tricycle landing gear, he descended from the heavens in the "death dives" that were his specialty. His brother Hillery, also an early dirigible and aeroplane pilot, recalls in *The American Heritage History of Flight*:

Lincoln's main stunt was to come down head first, to make a vertical dive from 5000 feet with his motor shut off and land exactly where he wanted to.

I remember once in Dallas, Texas, he made two landings from this dive of his, and each time he came down within a foot of the other. He used to dive onto a race track, and he'd come to the starting wire and dive under it, and not hit the ground or the wire either.

Beachey was widely known for a circus act that pitted his biplane against the 300-hp racing automobile of the begoggled, cigar-chewing Barney Oldfield for the "Championship of the Universe"; the aeroplane generally won. He gathered more laurels on June 27, 1911—flying over Niagara's spectacular Horseshoe Falls, under the steel arch of the International Bridge, and down the gorge in six minutes of supreme suspense for a hundred fifty thousand spectators—for which he received a cash award of $5000. He also accomplished the first indoor flight ever recorded—in the Palace of Machinery at San Francisco's Panama-Pacific International Exposition. With the example of Pégoud irresistibly before him, Beachey turned looping, upside-down flying, and vertical banks into signatures of his own by such embellishments as the "corkscrew flop," the "dip of death" (or vertical drop), the "Dutch roll," the "ocean roll," the "turkey trot," and the spiral and reverse. As he put it in a brochure:

The Silent Reaper of Souls and I shook hands. Thousands of times we have engaged in a race among the clouds—plunging headlong in breathless flight—diving and circling with awful speed through ethereal space. And, many times, when the dazzling sunlight has blinded my eyes and sudden darkness has numbed all my senses, I have imagined him close at my heels. On such occasions I have defied him, but in so doing have experienced fright which I cannot explain. Today the old fellow and I are pals.

At the same time, Beachey wrote poetically in his prospectus "The Genius of Aviation" (price one dime) of the quiet and peace experienced in flights across valleys, over mountains, and "high up among the clouds full of wondrous gold." "Will you not journey with me?" he asked rhetorically in the pamphlet. "I will not!" was the reply he fancied would be made by his readers. "Oh, very well . . . I will forgive you," he continued, ". . . and I will dare to say you have not nor will you experience life to its fullest until you have taken a trip to the clouds and skies." He topped the world record for such trips by breaking the altitude mark at Chicago in 1911, attaining 11,573 feet in his 60-hp Curtiss.

Beachey's extensive experiments with the laws of gravity showed what an aeroplane could do in the hands of a nervy and masterful pilot, thus adding to man's knowledge of aerodynamics and of structural strain under extreme conditions. And spectators were carried away by the sight of his soaring machine. "Each art has its master worker—its Paderewski, its Saint-Gaudens, its Michaelangelo, its Milton," said Elbert Hubbard in an effusive testimonial. "There is music and most inspiring grace and prettiest poesy in flight by man in the heavens. And posterity will write the name of Lincoln Beachey as the greatest artist on the aeroplane. In his flying is the same delicacy of touch, the same inspirational finesse of movement, the same developed genius of Paderewski and Milton."

Beachey himself was under no illusions that the crowds came solely to witness his skill in putting his special reinforced Curtiss Model D biplane through its paces in the skies. Mostly, he felt, they came to see him get killed—for people predicted that he would die in flight, and he made money on the bets they put up. For a short while he did renounce all flying, shocked by the deaths of those who tried to imitate him. Eventually he wrote his own epitaph: "If it came my time to bow to the scythe-wielder," he said in a newspaper article, "I wanted to drop from thousands of feet. I wanted the grandstands and the grounds to be packed with a huge, cheering mob, and the band must be crashing out the latest rag. And when the ambulance, or worse, hauled me away, I wanted them all to say as they filed out the gates, 'Well, Beachey was certainly flying some!'"

There seemed to be some characteristic common to those who dared death in the skies that repelled them from somber funerals. When Graham Gilmour, the brilliant young English flyer, was killed on February 17, 1912, while flying a new Antoinette-powered Martin-Handasyde monoplane from Hendon to Brooklands, he left instructions that the services should be "merry and bright" and that there should be "no mourning or moaning." In accordance with his request, few of those at the graveside wore black, and the masses of flowers were a jubilant riot of color.

The circumstances of Beachey's last flight came close to fulfilling his own melodramatic wishes. Fifty thousand people watched him take off from the polo grounds at the Panama-Pacific Exposition on the afternoon of March 14, 1915, piloting a light monoplane specially built for exhibition work by his friend Warren Eaton. The crowd held its breath as he started his frightening vertical drop from 3500 feet, with the power full on—the first time he had tried the stunt with this new machine. At 500 feet, as he tried to come out of the dive, both wings broke off with a loud report. Beachey crashed into San Francisco Bay and was drowned. His body was extricated from the wreckage thirty-five minutes later, while the crowd silently looked on. There seems little doubt that, with a windshield in his face for the first time instead of the breeze that blew freely around him in the Curtiss pusher, he misjudged the speed of his descent and pulled up too sharply and too late.

In an effort to get away from stunting by reviving the sporting aspects of aviation, the *New York Times* sponsored, on October 13, 1913, an aerial derby around Manhattan, in conjunction with a meet arranged by the Aeronautical Society of New York at Oakwood Heights, Staten Island. The Aeronautical Society, an offshoot of the Aero Club of America, had been formed because of the dissatisfaction of some members with the club's early emphasis on lighter-than-air activities—when the future obviously belonged to the aeroplane. Ten years had elapsed since the first flight of the Wright brothers; and the society thought it fitting that tribute be paid to the late Wilbur and the other "brave men who have passed over and who now sit at the table of gallant men." The *Times* went along with these praiseworthy sentiments to the extent of $2250 in cash—$1000 for first place, $750 for second, and $500 for third. Compared with the rich rewards and imposing goals set up in Europe, the derby around New York was little more than a sideshow. What made it remarkable, however, was that the race took place in a galloping north-westerly gale—a weather factor that probably would have caused its cancellation a year or so earlier.

The *Times* contest had no international aspect like the glamorous gathering at Belmont Park in 1910; but the newspaper was applauded for its attempt to rescue American aviation from the doldrums. And for one brief moment it managed to focus the eyes of the aeronautical world on Manhattan. In 1913 curiosity rather than enthusiasm marked the attitude of most New Yorkers toward an aerial derby—mingled with the feeling that if those fellows wanted to risk their necks, that was their business. The city itself was a unique setting for a race in the skies: the course was over water practically all the way, and the tall buildings provided convenient vantage points from which to watch the daredevils. Admission was free to the man in the street; unnumbered thousands could simply pause in their daily routine and tilt their heads to follow the flights. It would not of course be easy to identify the contestants, for their small machines would be mere specks against the clouds—but there was nothing to prevent partisan speculation as to which was which.

On Sunday, October 12, the *Times* printed the names of eighteen "entrants"—some of whom, it developed, had never heard of the contest. For this, and for "hasty and impromptu arrangements," it was roundly criticized by certain members of the aviation fraternity. The *Times,* however, was acting in good faith. Omitting the superstitious "13," the newspaper proceeded to assign numbers for the order of start, to be painted on the wings and tail of a miscellany of hastily groomed machines.

Several participants had flown to the starting point from Hempstead Plains, where (according to one reporter) "men walked around the field discussing the race and the types of aeroplanes that will take part in it, just as an English Derby crowd walks about the paddock and discusses the good points of the horses and jockeys." But when the starting gun was fired at 3:30 next afternoon, only five were poised for flight. Threatening skies and the blustery wind were formidable deterrents. Some hopefuls were frustrated by motor trouble, some by inability to assemble and test their machines on short notice, others by failure to obtain promised deliveries.

The course was clear enough: from Oakwood Heights around Manhattan and back again to Staten Island. For safety's sake, each flyer was required to wear a life preserver (an inflated inner tube would do) and to carry a map showing the route and the control points. As a further precaution, it was stipulated that the race be flown at an altitude of not less than 2000 feet, to allow for an emergency landing elsewhere than on the housetops of New York.

Moisant International Aviators, with three machines in the contest, ran an advertisement in the *Times* pronouncing themselves the "Largest Group Of Flyers In The World"; giving their motto as "At Home Among The Clouds"; and exhorting the public, "Watch our flyers in the sky." As if to underscore the claim, the number "1" had been allotted to C. Murvin Wood, pilot of a new Moisant Blue Bird monoplane. Wood

One of the first Americans to loop the loop, Charles Frank "Do Anything" Niles took second place in the 1913 "aerial derby" around Manhattan Island. He is shown here in his Moisant monoplane at Hempstead Plains, Long Island. (Photographed by the author.)

was regarded as a sure winner, and his machine as the best of the lot. In accordance with his assigned priority, he was the first to take off.

J. Guy Gilpatric, his head encased in a thick cork helmet, was No. 2 and the next pilot up. His machine, a newly tested Deperdussin, had been introduced into the United States and manufactured under license by the Sloane Aeroplane Company of New York. Because of the Deperdussin's reputation for speed, its supporters believed that Gilpatric's machine—built on conventional rather than *monocoque* lines—had the best chance to beat the Blue Bird. A Sloane advertisement alongside that of Moisant also bade New Yorkers to "watch our entries in the Times Aerial Derby." Gilpatric, a youthful adventurer in the air, was recognized as the main expert on this foreign plane in the United States.

Third to climb into the rough weather was William S. Luckey (No. 7), in a Curtiss pusher. This machine was substantially the same in basic design as that with which its inventor had won the first Gordon Bennett race in 1909; but its big Curtiss eight-cylinder motor, rated at 100 hp, was thought to be a telling factor in any contest for speed. Luckey was a middle-aged New York businessman who flew for sport and who was relatively unknown in professional aviation circles. Perched far out in front of the wings, he was completely unprotected from the wintry blasts. But his dexterous flying, combined with the proven qualities of the Curtiss (and, some said, the magic in his name), constituted a favorable prospect for him.

Charles Frank Niles (No. 8), a twenty-three-year-old native of Rochester, New York, in a machine identical in every respect, took off fourth and was close behind his fellow Curtiss flyer. "Do Anything" Niles was considered an exceptionally able and versatile pilot. Later he was to join the Moisant camp and become one of the first Americans to loop the loop and fly upside down.

Last to ascend was Antony Jannus—No. 9 on the list—who had shipped his Benoist biplane with 75-hp Roberts motor all the way from St. Louis for the contest. The Benoist was not a fast machine, but it flew surely and steadily in a wind; and the experienced Jannus, well known for his repeated assaults on American records, commanded plenty of respect.

A shivering herd of spectators—estimated at five thousand by the *Times* (at four thousand by the rival *Herald*)—had come to Oakwood Heights by touring car, by bicycle, and by the Staten Island ferry. As the fragile craft bounded across the field at one-minute intervals, rocking in the gusty air as they took off, the crowd yelled and waved Godspeed. Sirens screamed and tugboats let loose with their whistles as the flyers struck out across the bay; but the pilots were oblivious to everything save the task of keeping level in the treacherous wind. When the last plane had disappeared in the distance, the crowd settled back for an hour's wait. Their apprehension was alleviated only partially by the serenading of a brass band. "Through the Clouds"—a lively two-step picturing, on its sheet music, a young fellow and his girl riding a reasonable facsimile of a Wright biplane—had succeeded the catchy "Come, Josephine, in My Flying Machine," which everyone had been whistling a year or so earlier. But more appropriate to the weather that day was a repeat request for the popular "Bobbing Up and Down."

If the audience felt cold, the performers were congealed. Luckey, suffering from rheumatism and wearing only a sweater in addition to ordinary street clothes, could scarcely move his hands and feet, and clung to the steering wheel as if frozen to it. Niles fought the repeated frosting of his manifold pipe: soon after passing the junction of the Harlem and Hudson rivers, he began to "kick the throttle wide open and begin knocking off the frost every three or four minutes." No less a handicap was the high wind. At Spuyten Duyvil the turbulence gave Murvin Wood a sideslip that nearly caused him to turn over; before he could regain control, he had dropped three hundred feet. Gilpatric encountered the same peril. At one point the sudden impact of a strong gust caused his left wing to drop alarmingly, and from a height of three thousand feet he was forced to dive straight downward to recover. Jannus found the Battery the most dangerous spot; there the tall buildings deflected the wind, and he had to go over a jarring series of bumps, "exactly like riding a bucking horse." Someone else described it as feeling "like a leaf tossed in the autumn gale." Whatever the sensation, each aviator underwent an experience he would not soon forget.

One by one the birdmen returned safely to their roost, strung out by a difference of 21 minutes between the first arrival and the last. To the surprise of those who had counted on the monoplane for speed, a biplane was first—and second as well. Luckey, with Niles on his heels, shot over the

The sheet music for "Through the Clouds," the lively song played at the 1913 aerial derby, portrayed a young man and his girl riding in a Wright biplane, dressed in the flying fashion of the day.

finish line in his Curtiss well ahead of the pack. The best that Wood and the Blue Bird could do was to place third, more than 5 minutes behind the victor. Gilpatric and the Deperdussin were fourth; Jannus's Benoist was a slow fifth. For the approximately sixty miles around Manhattan, the time of each was as follows:

1.	Luckey	53 minutes 6 seconds
2.	Niles	55 minutes 5 seconds
3.	Wood	58 minutes 19 seconds
4.	Gilpatric	1 hour 9 minutes 6 seconds
5.	Jannus	1 hour 14 minutes 7 seconds

It was a chilled, exhausted, oil-covered five who were helped from their machines and into the shelter of the hangars. Drawn faces, numbed hands, eyes bloodshot behind goggles, testified to the strain. All agreed it was no day for an inexperienced flyer. "I wouldn't fly the race again under similar conditions for two thousand dollars," Luckey told the *Times*.

Since the objective of the Aeronautical Society was "to give credit to those whose sacrifices have brought the art of aviation to its present reasonably safe condition," there was cause for rejoicing that the race itself had been run without a single crash. The jubilation was dampened, however, by the prior loss of one of the entrants, who vanished literally into thin air en route to the scene from Hempstead Plains—a mystery that was never solved.

Albert S. Jewell, a mechanic who had obtained his license at the Moisant school six months before the race, took off from the Hempstead aerodrome at eight o'clock on the morning of October 13, to be on hand for the starting gun at Oakwood Heights that afternoon. His route lay over Jamaica, Jamaica Bay, and Coney Island, and across The Narrows to Staten Island—in sight of land all the way, and of countless pleasure craft that dotted the surface of the water. Mrs. Jewell and her three-year-old daughter had gone to Oakwood Heights to watch her husband compete; and until the last moment officials expected him to appear. The race could not, however, be delayed on Jewell's account, and was begun without him. When no word was received of his whereabouts, an intensive search was initiated by telephone, by wireless, by ship, by automobile—even by aeroplane. But all was to no avail.

When last seen, Jewell was over Belmont Park, speeding southward at a good height. Conflicting reports thereafter placed him off Far Rockaway and Rockaway Beach; and the presumption was that he had been swept out to sea by the wintry wind. For days hope persisted that he might have been picked up by a vessel lacking wireless. But on October 16 the flag of the Aeronautical Society at Oakwood Heights was placed at half-mast—in acknowledgment that one more life had been claimed by the advancing science of aviation.

While the *Times* derby may not have been the "greatest race ever seen in this country," as Jannus enthusiastically described it, the flyer was probably right in saying that no contest had "brought out so completely all that an aviator and all that his machine could stand up to." But from the standpoint of design innovation, the path of progress remained exactly as it had been—under the direction of European manufacturers.

CHAPTER TEN

Eccentrics and Geniuses

As soon as it had been demonstrated that man could "soar like an eagle," a host of fabricators in various parts of the world rushed to copy the proven models. If imitation is the sincerest form of flattery, then the Wrights, the Farmans, the Voisins, Blériot, and Levavasseur received more than their fair share of praise. Their successful designs were quick to influence others, and most subsequent models followed the lines established by them.

Much genius and energy, however, were expended in other directions. Scores of strange contraptions were built. Some of these became airborne, if only momentarily; a small minority achieved sustained flight; but many more—whose designers were ignorant of, or ignored, established rules of physics—never left the ground. Regrettably too, hundreds of misguided amateurs, whose machines were pathetically impractical, lost their life savings for their pains. Called "crazy" by their friends and neighbors, some actually did go out of their minds when the only flight to take place was that of their ideas and investment.

Scientific American, on June 5, 1909, felt constrained to editorialize on the subject:

When an age-long problem of such difficulty as that of the human mastery of flight is solved in a sudden and sensational manner . . . a stimulus is given to the art, the effect of which is seen in the immediate effort . . . to emulate if not surpass the achievement. Much of this endeavor, probably most of it, is doomed to failure, chiefly because the experimentalist does not realize the extreme difficulty of the problem, both from the theoretical and mechanical standpoint, and labors under the mistaken impression that a machine that is a broad imitation of the original must itself necessarily fly.

One of the world's great aeronautical engineers, Igor Sikorsky, recalled some sad facts in a 1964 lecture given at the Wings Club in New York:

Much of the design and flight knowledge that is now taken for granted was then unknown and . . . had to be learned through failures and tragedies . . . [there were many instances] where aircraft designed and built at tremendous effort, with mortgaging of properties and borrowing to the limit in the hope of achieving outstanding success, ended instead in total failures . . . The greatest failures and bitterest disappointments came from overconfidence and an uncontrolled imagination which the unfortunate inventor mistook for intuition. As a result, the inventor would disregard the study of former art, would not even try to verify his ideas with models or gliders.

Least likely to succeed, yet the most frequently attempted, was a design that imitated a bird—for the obvious reason that birds flew so well. The ornithopter, or flapping-wing machine, seemed wholly logical to dozens of learned scientists; even after the propeller-driven aeroplane was an accomplished fact, it claimed countless man-hours of study and experimentation. The dream was at least as old as the ancient Greeks, who told how Daedalus, the legendary inventor and architect, and his son, Icarus, escaped from the labyrinth of King Minos on Crete by fastening feathers together with thread and wax to make wings. Daedalus landed safely in Sicily; but Icarus, enchanted by the joy of flight, soared too high. The heat of the sun melted the wax in his wings, and (like some later birdmen) Icarus plunged into the sea. In more recent times another great inventor, Leonardo da Vinci, examined the same problem and proposed much the same answer. Among the legacy of drawings bequeathed by that many-sided genius was one of a mechanism intended to enable a man to flap a pair of wings attached to his body by moving his arms and legs. In the nineteenth century Sir George Cayley conducted intensive researches on bird flight. He built the world's first full-size glider, along the lines of the modern monoplane, and in 1853 launched the first manned (though not controlled) glider in history. And in France, Hureau de Villeneuve fashioned nearly three hundred artificial birds in varying shapes and sizes in order to study the secrets of flight, while Alphonse Pénaud created a model for an aerial device that flapped its wings to fly. (One of Pénaud's toys was to spark the curiosity of the Wrights, setting them on the path to their lofty achievement.)

Many others—including Louis Blériot—contributed to the theory of flapping flight and tried to realize in practice this seemingly simple way to leave the ground. But failure was their common reward. During the years 1907–1909, expectations in France centered on an ornithopter (the Collomb) that its backers insisted would take to the air; but it got little farther than the doors of its hangar. A dedicated constructor of ornithopters was Adhémar de la Hault of Belgium, who in 1907 put together a lightweight machine that represented the sum total of all efforts to date. Following his ideas, an Austrian named Soltau produced a birdlike contrivance in 1910. Unfortunately, both creations remained earthbound. Still in search of a solution, de la Hault completed a second machine in 1911: an orthodox-looking monoplane to which the ornithopter principle had been applied. But it too was an utter failure. A grotesque contraption with unwieldy, flexing wings was assembled by the German architect Scholtz in

The Dubois-Riant ornithopter dispensed with a propeller in favor of flapping wings that imitated the birds. But its wings beat the air in vain. It is shown here where it remained— on the ground.

1910—a monument to futility if only because it lacked the power to provide even the semblance of a lift.

Other ornithopter fanatics, European and American, continued to build flapping-wing machines in their back yards or barns—but the more they struggled to get into the air, the more firmly glued to the ground seemed their creations. One of the more serious of these efforts was the all-metal, 35-hp Dubois-Riant monoplane, which appeared in France in 1911. It was the claim of its inventors that ornithopter wings had a telling advantage from the military viewpoint: the absence of a propeller that could be shot away obviated the most vulnerable feature of an aeroplane in wartime! However, the wings of this relatively cumbersome machine also beat the air in vain; like others of its ilk, it accomplished little more than noise and waste motion.

Aside from the technical difficulties of achieving stability and control, a major drawback of all these flapping-wing craft was that the throbbing beat of the mechanism, in imitation of a bird's pinions, was likely to prove disastrous to the structure even before sufficient power was generated to lift the apparatus into the air. As the conventional aeroplane, with its aerodynamically sound lines, continued its rapid evolution, the ornithopter became, by degrees, of academic interest only; and it soon passed into the limbo of abortive flight experiments.

Equally unpromising at first, the helicopter attracted many aviation disciples who considered the ornithopter unworkable. To rise vertically from a starting point, instead of running along the ground for a hundred yards or more, was at once the hope and despair of an inventive French mechanic named Paul Cornu, scion of a family noted in the physical sciences. To construct a chassis that supported a motor was not difficult; but what to do about the downward currents of air created by the whirling horizontal blades was a puzzle that absorbed most of young Cornu's attention. He decided to check these drafts by interposing planes set at an angle to be determined by the operator. Glancing off the planes, the deflected currents would then supply the propulsion to drive the craft forward. Other technical problems presented themselves; and, as with most such projects, finances for the necessary research were in short supply.

Despite the obstacles, nonetheless, on November 13, 1907, at Lisieux, Cornu managed to lift himself and his machine a few centimeters—the first time that anyone had gotten off the earth using rotor blades instead of wings. For a power plant he used a 24-hp Antoinette motor; this was mounted on four bicycle wheels and drove twin twenty-foot propellers supported on outriggers. On a second occasion Cornu ascended with his brother—a total weight of 723 pounds—to a height of 5 feet, reportedly staying aloft for 1 minute.

Cornu's contemporary, the French engineer Léger, who worked under the auspices of Prince Albert of Monaco, conceived the idea of two propellers of contrary pitch revolving in opposite directions and mounted on two concentric axes for lift, while the common axis could be inclined for horizontal propulsion. His machine, however, never got beyond an earthbound testing stage. Another French device intended for vertical takeoff was the Vuitten-Huber, an elaborate rig of counterrevolving rotors driven by an eight-cylinder, 45-hp motor; it, too, was never airborne.

In the United States, a counterpart of the Vuitten-Huber was built in 1908 by G. Newton Williams, who made use of the 40-hp, air-cooled V-8 engine developed by Glenn Curtiss for the Aerial Experiment Association—but it failed to fly. Emile Berliner, member of a well-known American family of inventors, became imbued with the conviction that the Wrights were wrong and that the only way to fly was straight upward; but for many years he had no more luck than Williams. Not until after World War I did Berliner and his son, Henry, develop several moderately successful prototype helicopters—the best of which could achieve heights of 20 to 25 feet.

It is to another French inventor that the lion's share of credit belongs for improvements in helicopter design. Louis Charles Bréguet, with the assistance of his brother Jacques and of Charles Richet, began experimenting in June 1906 with a "gyroplane," which combined the fixed surfaces of an aeroplane with the revolving blades of a helicopter. On September 29, 1907, at Douai, this machine—supported by four poles for stability—allegedly raised a man off the ground vertically for about 2 minutes at a height of something less than 2 meters. The following year, on July 22, 1908, Bréguet is said to have negotiated a distance of 20 meters at double the original height.

The Bréguet-Richet gyroplane, with its four big propellers, was followed by a much larger "convertiplane" in 1909. It was not a success. Baffled by the apparently insoluble problem of helicopter stability, Bréguet gave up and turned to better known designs. His solidly constructed tractor biplane, the Bréguet IV (nicknamed the "Coffee Pot" because of its appearance), made a world record by carrying six persons in August 1910. The machine developed steadily into one of the most stable and rugged of pre–World War I aircraft; so adaptable was it to military use that it became a favorite with the French and other armies. But a successful helicopter (or "autogyro") was not to appear until after the war.

In the course of the early years practically every conceivable form of wing surface was tried—and generally found wanting. To mention only a few, there were, for example, such bizarre concoctions as the circular cage of the Frenchman d'Equivelly, containing five small surfaces on each side of the motor, with two larger surfaces on top; the double-celled, tunnellike "wings" connected by fuselage built by the Vermorel company for another French inventor, Givandin; and the arrangement of a couple of wings placed lengthwise instead of crosswise, with a propeller both fore and aft, produced in 1910 by Dr. Dane Hulbert, a Swiss. Needless to say, such unconventional efforts got nowhere. The earliest product of de Pischoff and Koechlin, a tireless team of experimenters in France, was on the order of an elongated box kite, in which the pilot lay prone, and with a tractor propeller not much larger than an electric fan. But like other abortive designs of the period, it, too, lacked the capacity for sustained flight.

Then there was the tandem monoplane of the Austrian Hipssich: one wing behind the other, with propellers fore and aft of the rear wing. Ungainly and unmanageable, the tandem plan was copied by the French engineer Guilbaut at Rouen in 1911; but his cumbrous flying machine was underpowered, with a motor of only 12 hp. Embarking on another tack, the designers of the Picat-Dubreuil tractor monoplane abandoned rigid wing surfaces in favor of pliant, supple sailcloth, rigged to a semicircular wooden spar that served as leading edge. Although otherwise conventional in form, this French conception lacked any provision for lateral control—apparently relying on hope to navigate the air safely. There is no record of its having ever flown. In England, the Walton Edwards Rhomboidal, which also departed from rigid lift surfaces to employ an arrangement of horizontal sails, remained resolutely attached to the soil—while the Lee Edwards annular biplane, like a doughnut in design, gave little evidence of possessing any flying qualities save a certain inherent stability.

In those feverish early days of experimentation, men would do anything to get into the air—even if for a few seconds; even without any means of power save their leg muscles; and even if they made spectacles of themselves in the process. Thus was born the *aviette:* a bicycle built with wings and tail, with or without propellers, the power for which was furnished by furious pedaling. Some *aviettes* actually "took off"; many more reared up and crashed, or hugged the earth regardless of the "pilot's" efforts. But behind the seriocomic contests periodically held for them in France was the thought that some new wing form or flight principle might possibly emerge. *Aviette* contests were a sort of poor man's aviation tournament; the participants were happy if their kitelike contrivances generated enough lift to rise a few inches and "leap" a few feet without losing equilibrium.

While disenchantment came quickly to many offbeat inventors who thought that they too could fly, it is remarkable how some ideas persisted longer than others. That of the tail-first aeroplane—the *canard*—is one

Bicycle-powered kites known as aviettes enjoyed a brief popularity in serio-comic tournaments in France. Some of these contrivances actually "flew" for short distances.

example. It was an important stimulus in the early work of both Blériot and Gabriel Voisin—the latter being among the few who applied it successfully. For military purposes, an all-metal monoplane of the *canard* type was devised in France by Lieutenant Biard and equipped with a 50-hp Gnôme. A steel-and-aluminum version in biplane form, mounted on a four-wheel undercarriage and powered by a 60-hp Anzani motor, was tried out by Captain Morel. Marcel Besson built a high-wing, heavy amphibian along the same lines, its 50-hp Clément-Bayard motor placed above the fuselage and to the rear of the cockpit; stability was obtained by ailerons and fins at the wing extremities.

Since weight was a dominant factor in designs of that era, when engines were too weak to lift a complicated structure, it was virtually impossible for such oddities to perform satisfactorily. Even if equipped with what was then a respectable power plant, a machine that deviated too far from orthodox form was foreordained to failure. The broad-winged Arnoux monoplane, tested at Issy by the Voisin pilot Louis Gaudart, had a Chenu motor rated at 59 hp—but possessed neither tail nor fuselage, whether front or rear. A contemporary effort in England, the spidery, lightweight, tail-forward Valkyrie monoplane (completed in February 1910), was much more successful.

Some of the oddest-looking flying machines, ridiculed as freaks when they made their appearance, actually anticipated modern designs. One example was the British Blackburn monoplane of 1909, in which the pilot and engine were suspended beneath the fuselage—foreshadowing the "parasol" plane developed early in World War I to provide greater visibility on reconnaissance. (The true progenitor of the parasol machine, however, was the 1907 model of Santos-Dumont rather than the Blackburn.) And the "delta wing" concept was anticipated in England about 1910 by a Captain Wyndham, whose machine featured a large triangular

Anticipating in appearance the modern plane with retractable wings, the Marçay-Moonen monoplane of 1912 was constructed with wings that swung back to facilitate transport and storage.

wing and a smaller triangular tail. Wyndham's plane, however, did not fly and hence had no practical influence on design. Another early novelty—a hit at the 1911 Paris Air Salon—was the two-seater, folding-wing Marçay-Moonen monoplane, which was designed to travel on land as well as in the air. Powered by a 50-hp Gnôme, the Marçay-Moonen was equipped with two sets of controls. The first set enabled the pilot to swing back the wings along the sides of the fuselage and to steer two small wheels positioned near the tail; the second set of controls governed normal operations in the air. The folding wings were designed to facilitate stowage and transport rather than to improve the aerodynamic qualities of the plane; in this respect the Marcay-Moonen anticipated the aircraft-carrier planes of World War II rather than the F-111 swing-wing fighter-bomber of the 1960's.

It was inevitable that some constructors should copy the graceful arched wings of a bird in flight—in fixed, not flexible, form. Such a type was seen on the Franz Miller monoplane—the first machine to be built in Italy—which flew, after a fashion, in 1908 and 1909. The frigate bird was the model for a French tractor monoplane with unmistakable Blériot characteristics, powered by a three-cylinder, 30-hp Anzani motor. Named, naturally enough, La Frégate, it was tested at Issy by Robert de Lesseps. José Weiss's monoplane in England, the product of extensive studies of birds on the wing, likewise had arched wings, which were immovable rather than of ornithopter style. In Germany the elegant, high-curved wings of Dr. Geest's monoplane, the Möwe ("Sea Gull") were frequently seen during the early days at Johannisthal; and the twin-propellered Loose monoplane in the United States was an experimental step in the same direction.

But those who studied the straight-outstretched, motionless wings of birds like the condor, the hawk, or the vulture—which can swoop and glide for hours—were closest to the right solution. Best known of these birdlike

Several designers patterned their fixed-wing, inflexible constructions after the arched shape of a bird in flight, as shown here.

The Péan monoplane

Robert de Lesseps's monoplane, La Frégate

The Guilbaut monoplane

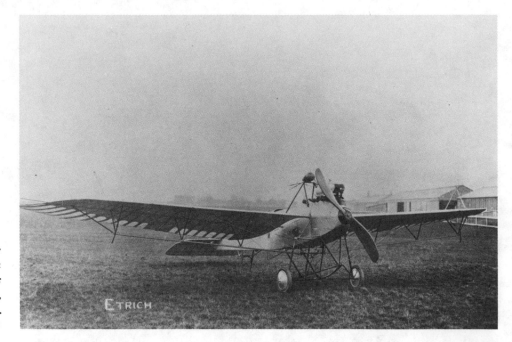

The Etrich Taube monoplane, an Austrian design, was named for its long, dovelike wings. (National Archives)

models was the Taube ("Dove"), developed by the Austrian engineer Wels in 1908, in conjunction with Igo Etrich. It led to a whole family of related "pigeons," "gulls," "swallows," and the like—until designers discovered the effective modern form of wing structure.

One workable type was the triplane, which seemed to many a logical extension of the biplane: if two surfaces could give so much lift, an extra wing, supposedly, would give that much more. Such was the concept of A. V. Roe, whose tractor triplane made "flights" of up to 300 yards at Britain's Lea Marshes in 1909. The machine took off with a two-cylinder J.A.P. engine of only 9 hp—the lowest horsepower to be utilized in Europe—which illustrated a very early use of mechanically operated overhead valves. Roe's second triplane is preserved in the Science Museum in London.

The triplane in an experimental form was seen at an early date in the United States, for Octave Chanute made frequent use of a three-surfaced glider in his researches near Chicago. One of the first American triplanes to be fitted with a motor was produced by Morris Bokor in 1909. For stability, it had a pendulum seat like that of the Danish engineer Ellehammer; for power, it had a four-cylinder A&B automobile engine weighing 419 pounds, rated at 38 hp. Between the lowest and middle wings was a gap of five feet; between the middle and uppermost, a gap of six feet. This arrangement was in the belief that somehow the two eight-foot pusher propellers would draw the air back below the middle wing and thus tend to "neutralize" any interference of the bottom wing. Like the Wright biplane, the Bokor model had a forward horizontal rudder, as well as a double tail of triangular panels. With the pilot in his seat, it weighed 1181 pounds. Bokor's theories were never proven; the craft lacked sufficient power to get off the ground.

In Europe the triplane met with more extensive recognition. Hans Grade's tractor triplane was the first German-built and German-piloted plane to fly (in 1909)—although the builder soon discarded it for a hybrid monoplane that combined the characteristics of the Blériot, Antoinette, and Demoiselle. In Switzerland, the first homemade machine to fly was likewise a three-decker—a dihedral affair built by the Dufaux brothers in 1908. Ambroise Goupy, the French pioneer, began with a cellular-type triplane constructed for him by Voisin Frères in 1908; however, he later changed over to a tractor biplane with staggered wings that had much influence on European aviation. In 1911 the Société Astra—licensed to use the Wright wing-warping system—succeeded in putting into the air a triplane which, despite its excessive weight, turned in a noteworthy performance at the hands of Marcel Goffin, an early Antoinette pilot. For a framework the Astra used strong steel tubing; the body was enclosed with metal sheeting, and six extra-heavy rubber-tired disk wheels (two at the rear) minimized landing shocks. With its Renault motor of 75 hp, it could carry a useful load of around 880 pounds. (Empty it weighed 1650 pounds.) The Astra company also later dropped the triplane for a biplane, which was not only more economical but gave better results in both land and water operations.

Among other triplanes was the one constructed by the talented Louis Paulhan: a pusher entirely of steel tubing, with a short nacelle suspended between the middle and lowest wings and with a four-wheeled undercarriage. Also powered by a 75-hp Renault motor but more than a hundred pounds lighter than the Astra, it was designed to carry two passengers in addition to the pilot. Paulhan constructed this craft for the French military trials in 1912, but it failed to make a significant impression. At about the same time, seeking new forms and formulas, Paulhan produced the Aéro-Torpille—a monoplane with upcurved wing tips and a shaft-driven propeller at the tail. The "torpedo," however, though highly advanced for its day, did not survive long in practice.

The French Société Astra triplane had a body encased in metal sheeting and six rubber-tired disk wheel. to support its heavy weight on the ground. In spite of a notable performance by its pilot, Marcel Goffin, the design proved impractical.

The Paulhan triplane, built and piloted by Louis Paulhan, was driven by a propeller mounted in the rear and powered by a 75-hp Renault motor. It was designed for military trials.

The idea of additional wing surfaces in the interest of lift did not of course stop with the triplane. Matthew B. Sellers of Baltimore had a try at a quadriplane. His aim was to fly with the least possible horsepower; thus he experimented exclusively with small, light power plants. In 1908 he constructed a flying machine with four wings staggered one above the other, each measuring eighteen by three feet; he also designed and built an air-cooled two-cylinder engine that weighed only 23 pounds and delivered 4 hp. Total weight of plane and motor was 110 pounds. Sellers began by making short, straightaway hops; later, with a second engine of about the same weight, he was able to climb, turn, and fly at about 21 m/hr. Sellers contributed a number of valuable studies to the science of aeronautics. But his quadriplane was overtaken by the rapid development of monoplanes and biplanes, with their stronger and more efficient structures.

The next step, naturally, was a machine with five wings. Professor Jerome S. Zerbe of Los Angeles put together in 1909 an apparatus that had five double-cambered wings (and afterward one with six wings)—but without ever finding a power plant adequate to achieve flight. At the Johannisthal aerodrome a year or so later, the German constructor J. Merx aroused much curiosity with a staggered *fünfdecker* ("five-decker"), which had two chain-driven propellers and a motor mounted on a four-wheeled platform; but like most such multiple-winged craft, it lacked the elements for successful flight.

While many early inventors grasped the principles of wing design, one of the most common causes of failure was inability to master the problem of lateral control. Stability in a fore-and-aft direction was relatively easy to attain with the proper use of elevators; but endless controversy and experimentation were expended over the merits of wing warping as opposed to hinged ailerons to maintain balance laterally. Still another school of thought chose to depend on the "inherently stable" characteristic of the machine to implement its flight through the air without the aid of ailerons, wing warping, or some other device to keep it on an even keel.

Standing out in welcome contrast to a depressingly long list of failures was the biplane built by Dr. William Whitney Christmas of Washington, D.C., which was controlled laterally—as well as partially steered—by interconnected ailerons on the outer trailing edge of the wings, after the fashion of Curtiss's models. The Christmas biplane made a straightaway flight at Fairfax Courthouse, Virginia, as early as 1908—the proof of which became of considerable cash value to its inventor later on. His patent application, covering recessed ailerons, was not granted until May 5, 1914; but on the strength of this pioneer performance, the Christmas Airplane Company brought a claim against the U.S. Government for infringement. In 1923 the Government agreed to buy the patent for $100,000—a fair enough return on "Doc" Christmas's original investment.

Not all were so fortunate. Wilbur R. Kimball of New York, at one time the secretary of the Aeronautical Society and an adherent of the helicopter theory, exhibited in 1906 a rubber-driven model that had two "air screws," each fifteen inches in diameter, mounted on wheels; it weighed altogether about ten ounces. According to a 1907 publication of the Aero Club of America, it could run 12 feet along the floor, rise, and fly for 70 feet. Kimball went on, under inspiration of the Wrights' success, to unveil a biplane in July 1909—the first machine constructed under the Aeronautical Society's aegis, and the first in the world to attempt the use of six four-bladed propellers. Anna Held, the reigning favorite of the New York stage, was persuaded to christen the Kimball creation at the society's flying ground in Morris Park. It was good publicity: the vivacious actress was singing a song about an aeroplane in a current Broadway

Two steps beyond the triplane, the five-decker Merx multiplane was never able to achieve sustained flight in its trials at Johannisthal.

show, and the ceremony drew a crowd of photographers. Sadly, what followed was a theatrical anticlimax—for Kimball was unable to get his machine into the air.

Others took failure to heart, with tragic results. Lieutenant Alexander L. Pfitzner, a former artillery officer in the Hungarian army who had emigrated to the United States, was induced by Glenn Curtiss in 1909 to come to Hammondsport to assist in the development of Curtiss aeroplane engines. Pfitzner soon became absorbed in the twin problems of control and propulsion as relating to flying machines. The difficulty presented by the Wright patents weighed heavily on his mind. In order to avoid infringement, he worked out an intricate arrangement of transverse, or lateral, controls for a new type of monoplane. The principle embodied in his novel machine was that of sliding surface extensions, or "equalizers," installed in the wing tips, to provide variable wing area control such as used by birds in flight. The increased extension of one tip caused greater lift on that side and hence a tendency to turn in the opposite direction.

The Pfitzner monoplane—a maze of wires and struts, with a four-wheeled undercarriage—was tested by the inventor at Hammondsport in 1910, but crashed after half a dozen flights. Pfitzner subsequently went to work for the Burgess Company at Marblehead. There he designed a biplane that incorporated the sliding-wing-tip principle; but it too was demolished in an accident. Thoroughly depressed, the Hungarian rowed out into Marblehead Harbor with his aeronautical drawings in a suitcase. Neither he nor his drawings were ever seen again; but a pistol, recently fired, was found in the empty boat. There was no doubt that Pfitzner had committed suicide.

Another melancholy note was struck by the happenstance of acute financial distress caused the survivors of many who gave their lives exploring novel aerodynamic theories. One story is illustrative—that of Frank E. Boland of New York, who also tried to avoid infringement of the Wright patents. At his workshop in Newark, Boland produced a tailless biplane, which first flew in 1911. Powered by an eight-cylinder, 60-hp motor built by the inventor, it had a horizontal rudder on a forward outrigger and depended for both steering and lateral stability on two hinged vertical triangular flaps, or "jibs," one at each extremity of the wings. "The Boland Control is the embodiment of utmost safety and simplicity in a new system of control which is basic in principle" was the company's confident assertion; yet when Boland incorporated his new system in a newly developed flying boat, the result was a crash that killed the designer, at Port-of-Spain, Trinidad, on January 23, 1913. The following message in the personal columns of the August 30 *Aero & Hydro,* seven months later, might have spoken for more than one bereaved mate of a pioneer:

Aviator's Widow Appeal—Mrs. Frank Boland must sell half interest of Boland Aeroplane and Motor Company at sacrifice. She was left penniless by husband's death. Investigate. Help widow who can't get a dollar from her late husband's aeroplane patented inventions.

As it has in the case of various other scientific inventions, the Soviet Union has appropriated the credit for inventing the aeroplane. While claims without number and without proof have been adduced in behalf of many experimenters (most notably by the admirers of the French pioneer Clément Ader), the Wright brothers have been recognized by consensus as the first of mankind to fly in a heavier-than-air craft under power and control. The Russians, however, waited until 1949 to assert the contrary. On July 20, 1882 (so said the Soviets), Alexander Feodorovich Mozhaisky astounded St. Petersburg by making a successful trial flight in a steam-powered flying machine—a whole generation ahead of the Wrights. The *Bolshoi Encyclopedia* shows a blueprint of Mozhaisky's broad-wing monoplane, which the Communists credit with impressively modern and ingenious features. It had ailerons, a tail unit consisting of elevator and rudder, a fabric-covered fuselage, a four-wheeled undercarriage fitted with shock absorbers, a compass, a bank indicator, and two engines (designed by Mozhaisky himself) that delivered 30 hp each. One engine was mounted forward with a tractor propeller, the other in the rear—to drive two propellers functioning through slits in the wing.

The fact seems to be that a patent for "an aeronautical device" was granted to Mozhaisky by the Czarist government in 1881; the Czarist *Military Encyclopedia* of 1914 describes a "test flight" made in 1884, indicating that the apparatus fell to one side and smashed a wing after briefly breaking contact with the ground. The Soviets, however, quote eyewitnesses to the "flight"; cite yearbooks and newspapers for three years after the event; and insist that documents reposing in Leningrad archives incontestably prove that the Russian inventor's work preceded anything done by Wilbur and Orville Wright. A Soviet book published in 1949 emotionally referred to Mozhaisky as "the majestic figure standing at the head of the glorious Pleiades of Russian aviation designers and brilliantly characterizing the originality and might of patriotic aviation thought."

The Communist claim for Mozhaisky was meticulously examined by Captain Glenn E. Wasson of the U.S. Air Force in the July 1962 issue of *Airpower Historian,* the journal of the Air Force Historical Foundation. Because access to official Soviet sources was lacking to Western scholars, Captain Wasson concluded that until this situation is remedied, the latter-day discovery of Russian airmanship should be taken with a liberal sprinkling of salt. (It is only fair to add that Russians of the post-Stalin period have not been so extravagant in their claims as to the perpetrator of the world's first flight.)

Whatever may have been the contribution of Mozhaisky to "patriotic aviation thought," the dream of flight was realized by another Russian, Igor Ivanovich Sikorsky, in infinitely more substantial form. Sikorsky was born in Kiev in 1889. After completing a general course at the Naval Academy in St. Petersburg, young Sikorsky read about Wilbur Wright's successful demonstrations at Auvours. Inspired like so many others by their pathfinding exploits, he quickly gravitated to the flying fields around Paris, where he began to absorb the yeasty atmosphere (if not the experience) of aerial activity. Against the advice of Captain Ferber, Sikorsky proceeded to draw up plans for a helicopter, meanwhile acquiring a 25-hp Anzani motor. Back in Kiev in the summer of 1909, he began his experiments in vertical flight—which, like all similar efforts along this unorthodox line, ended in failure. (They were not to be resumed until many years later.) At the same time he designed and built a small pusher biplane— Model S-1—powered by another Anzani of 15 hp that he had used for experiments with a propeller-driven sleigh. Attempts to get the S-1 into the air were not successful; but, using the 25-hp motor, his Model S-2 tractor became briefly airborne on June 3, 1910. Sikorsky wrote in his autobiography, *The Story of the Winged-S:*

> The S-2 was wheeled out of the hangar. I checked the controls, climbed into the seat behind the motor and shouted "contact!" The engine was quickly started. While three men held the plane, I gradually opened the throttle, and a few minutes later the good sound of the motor, the propeller blast and the smell of burned castor oil, told me that it was time to try. I gave the signal, and the plane was released. The S-2 had a much better acceleration. From the very first moment I could feel the stronger propeller blasts made the control more effective, and the tail went up at once. . . .

As in the case of all fledgling flyers, the inventor of the S-2 felt the inexpressible thrill associated with actual, if momentary, flight. And as customary in that school of elementary education, where most of the students were self-taught, the pilot cracked up. A slow-soaring plane, the S-2 did not have enough power to pull out of an air pocket unexpectedly encountered over a ravine. It was soon followed by Model S-3, with a 40-hp Anzani, which flew better but also crashed, on the frozen surface of a lake, on December 13, 1910. Not until April 1911 did Model S-5, with a 50-hp, water-cooled Argus motor, bring the satisfaction of complete success to its pilot. With this machine Sikorsky was granted a license from the Imperial Aero Club on August 18, 1911.

Sikorsky soon turned to the study of bigger and heavier aeroplanes. The S-5 had made a forced landing one day when a mosquito that had fallen into the gasoline was sucked into the carburetor, thus effectively stopping the engine. Ruminating on the constant risk of such engine failure, as well as on the problem of reduced speed caused by drag, Sikorsky decided to build a large three-seater biplane, powered by a 100-hp Argus motor—Model S-6. An improved version, Model S-6A, had an increased upper wing span and a wood-veneer fuselage enclosing seats for the pilot and two passengers and the engine. The rate of climb and the lifting capacity of this machine were so much better, and the streamlining so effective, that it surpassed the world speed record for pilot and two passengers by attaining 113 km/hr. Model S-6B was a two-seater biplane resembling the S-6A, but still further refined.

Following these encouraging results, Sikorsky began thinking, in December 1911, of even larger planes, with as many as four motors of 100 hp each. He entered into an agreement with the Russian Baltic Railroad

Car Factory, which gave him the position of designer and chief engineer of the company's aircraft division; this led to his moving to St. Petersburg in the spring of 1912. There he rented a small building and started work on his flying machines with the help of a staff that had grown to thirty men by midsummer. The chairman of the board of Russian Baltic, Michael Vladimirovich Shidlowsky, gave Sikorsky full support in production of the unconventional, giant machines that were to be his hallmark in the aeronautical world.

From September 3 to 26, 1912, the first Russian military competition took place at St. Petersburg, with eleven entrants. Sikorsky's plane, the S-6A, piloted by A. E. Raevsky and fitted with a 50-hp Gnôme motor, won the first prize of 30,000 rubles. Second place went to a biplane produced by the Fabrica Moscovita Tneerskaja—the Dux, flown by A. V. Gaber-Vlinsky. The third prize was won by a Dux monoplane piloted by E. de Boutmy. The trials were a pleasant surprise for the advocates of Russian aviation, which had been both lacking in original designs and slow to copy foreign models.

As soon as the competition was over, Sikorsky—now established as the leading Russian aircraft designer—began to devote all his energies to building the world's first great multiengine aeroplane. He had conceived of the "air giants of the future" as having, of necessity, "several motors, one independent of the other, as the only protection from hazards of forced landing"; he also had considered "the importance of a crew of several men who would assist and relieve each other and fulfill the duties of the pilot at the control wheel, the navigator, the mechanic and others. To enable them to perform their duties properly, a large, comfortable closed cabin was necessary, particularly in the severe climate of Russia."

Known as Le Grand, Sikorsky's huge biplane was far ahead of its time. Powered by four 100-hp, four-cylinder, water-cooled Argus motors, it weighed over four tons. With its wing span of ninety-two feet and its fully enclosed passenger cabin and portholes, it looked like something out of a Jules Verne fantasy. In front was a large balcony with a door leading to the pilot's compartment and its dual controls; from here entrance was gained to the luxuriously appointed passenger accommodations—four seats, sofa, and table. In the rear were a washroom and a closet. The row of windows gave excellent visibility, and there was full standing room everywhere.

This imaginative forerunner of the modern airliner was a radical and unique concept. Many learned opinions were advanced as to why it would never fly: it was too big and too heavy to rise from the ground; its multiple power plant would only create a dangerous imbalance if one motor failed; its size would prevent proper handling in the air or on landing; its closed cabin would render control impracticable because the pilot would not be able to feel the airstream on his face; and so forth. Yet fly it did, on the evening of May 13, 1913—to the joy and technical satisfaction of pilot Sikorsky, copilot Captain Gleb V. Aleknovich, and a mechanic

Using the facilities of a railroad-car factory, Russian aeroplane constructor Igor Sikorsky designed and built large, multi-motored planes with roomy compartments for pilots and passengers. His Ilia Mourometz set a world record in 1914 by carrying sixteen passengers. It is shown here coming in for a landing with two men standing on the upper platform. (I. I. Sikorsky)

waving happily from the bridge to an excited crowd below. On landing, Le Grand was engulfed in a roaring tide of thousands of spectators, who snatched up the members of the crew before they could properly step down from the plane and carried them off shoulder high.

The trial flight had confirmed practically all the claims that had been made for the mammoth craft and its many motors. After being inspected by Czar Nicholas II and completing fifty-three flights without trouble, Le Grand met a bizarre end. During the 1913 military trials, held from September 2 to October 14, Gaber-Vlinsky lost the engine in his pusher biplane and crashed into a ditch from a height of a thousand feet. The pilot walked away unhurt; but his engine had dropped on a wing of Le Grand, smashing it beyond repair. The loss was compensated for, to a large extent, by two of Sikorsky's pilots—Aleknovich and George A. Jankovsky—who flew the Model S-10 biplane and the Model S-11 monoplane, respectively, to victory in first and second places that year. Third place was taken by the Frenchman L. Janoir, in a Deperdussin, and fourth by the Swiss flyer Edmond Audemars, in a Morane-Saulnier.

Successor to Le Grand was the even bigger and better Ilia Mourometz—named for a legendary Russian hero of the tenth century—which was completed at the end of December 1913. Also incredibly advanced for its day, it had roomy compartments for pilots and passengers; a private cabin with berth, table, and cabinet; and, in addition to the front balcony, a stairway to an upper bridge. The plane was lighted by electricity, the current supplied by a wind-driven generator; heat was provided by two long steel tubes through which passed part of the exhaust. This machine made a new world record, on February 11, 1914, by carrying sixteen passengers. It was followed by the Ilia Mourometz II, with inboard motors of 140 hp and outboard motors of 125 hp—which on June 18 set a world duration record for pilot and six passengers of 6 hours 33 minutes. Late in June, Sikorsky took the machine on a historic round-trip flight of 1600 miles, with several stops, to his native city of Kiev, demonstrating once and for all the practicality of large, multimotored planes. The flight immediately elicited an order from the Russian army for ten machines of the same type.

But the shadows of war were falling fast over private industry. The great planes were in general too slow and too heavy for combat—although several of the ponderous craft did distinguish themselves in action. To the loss of Sikorsky's personnel under fire or in accidents was added the tragedy of the Revolution. Recalled to active duty by the Czar at the age of sixty, as chief of a newly created "Squadron of the Flying Ships," Shidlowsky was shot, together with his son, by the Bolsheviks in 1918. Sikorsky made his way abroad to New York—where he was one day to produce the famous amphibians, helicopters, and other craft bearing the emblem of the "winged-S."

CHAPTER ELEVEN

War Clouds

Prophecy

The time will come, when thou
shalt lift thine eyes
To watch a long-drawn battle
in the skies,
While aged peasants, too
amazed for words,
Stare at the flying fleets of
wond'rous birds.
England, so long the mistress of
the sea,
Where wind and waves confess her
sovereignty,
Her ancient triumphs yet on
high shall bear,
And reign, the sovereign
of the conquered air.

Thomas Gray, the English poet from whose *Luna Habitabilis* this stanza was taken, must have had an uncanny imagination—or the gift of second sight—to have prophesied "a long-drawn battle in the skies" and "flying fleets of wond'rous birds." The verses were composed at Cambridge in 1737—forty-six years before the invention of the hot-air balloon. The lines were lent poignancy when war clouds began to darken the skies of Europe in 1914; for England, though long the mistress of the sea, was far from being sovereign of the air.

When the fateful year opened, aviation had settled into a time of orderly development both in America and abroad. The great races and tournaments were over; the ever-mounting accident rate had, to a considerable extent, cooled public enthusiasm for the flying machine as the transport of the future; and while records continued to be regularly made and broken, that was no longer cause for surprise. Still regarded as a scientific "toy" or a plaything for the adventurous, without economic value to speak of, the aeroplane was yet to be proven as an important factor in military operations.

From adhering to Ferber's counsel of "step by step, jump by jump, flight by flight," aviation had steadily advanced through his next stipulated stages: "from ridge to ridge"; "from city to city"; and even "from continent to continent." Only one further stride remained ahead—across the oceans.

In 1913, at the height of the record-setting fever, Lord Northcliffe of the *Daily Mail* had fired imaginations by offering a prize of £10,000 to the aviator who made the first transatlantic crossing. To many, this appeared an unrealistic proposal; yet, with speeds of better than 2 miles a minute already attained, others looked upon it as within the realm of feasibility. Under the original terms of the offer, a flyer would be permitted to alight at sea to replenish his fuel supply. However, none of those who indicated they might try the long hop envisaged landings on and takeoffs from the water. From St. Johns, Newfoundland, to the coast of Ireland was approximately 1900 miles; the much longer southern route—via the Azores to a terminus at Vigo, Spain—was also considered, because of the break afforded en route.

Gustav Hamel, one of the most proficient flyers of the day, announced that he expected to accomplish the ocean flight in 1914 in no more than 16 hours. But before he could make the attempt, he set an ominous example for others who contemplated the feat.

Hamel had engaged in numerous cross-country and cross-Channel trips, commuting regularly between London and Paris. On April 17, 1914, in fact, he established a new kind of record for Britain along these lines. Although Europe was moving toward the brink of war, few people took seriously the question of defense against aerial attack. A small group of public-spirited individuals, however, had formed in February 1913, an "Imperial Air Fleet Committee"; in conjunction with the London *Standard,* they purchased by public subscription a two-seater, military-type Blériot with an 80-hp Gnôme motor, to rouse the country "to a realization of what other nations are already equipped to do, not only with one machine but with many hundreds." Under the committee's auspices Hamel used this plane to make a nonstop flight into the heart of Germany—from Dover to Cologne, a distance of 340 miles. Transporting a passenger, Hamel carried a total flying weight of 1729 pounds across the Channel and over five frontiers in 4 hours 18 minutes, encountering rain and hail throughout the journey but "without incurring the slightest damage to the machine or engine, whilst sufficient fuel remained to ensure a safe return to neutral ground."

Shortly after this remarkable performance, Hamel met Roland Garros in a match of virtuoso flying at Juvisy, where he won the high regard of his French colleague both as a flyer and as a person. Hamel then set out on the return trip to England, heading directly across the Channel. He

The first aeroplane purchased by any government, this 1909 Wright Flyer was catapulted into the air from a monorail at Fort Myer, Virginia. To move it along the track, a weight was dropped from the derrick in the background. (Air Force Museum)

was never heard from again. One may assume that he miscalculated his drift and, his gasoline exhausted, had found a watery grave. It did not bode well for a flight across the Atlantic.

Adolphe Pégoud then threw his hat into the ring, basing his hope of survival in case of forced descent on an escort of torpedo boats. Brindejonc des Moulinais and Marc Pourpe, champions of the world in long-distance flying, were expected to enter the lists. Roland Garros was mentioned as a likely competitor; and in Germany his friend and famous colleague Hellmuth Hirth proposed the construction of a six-engine hydro-aeroplane for his own attempt. In the United States, the New York *Evening World* announced that Harry N. Atwood planned to bridge the gap from Newfoundland to the Irish coast in 30 hours with his mechanic in an "improved hydroaeroplane." And the English flyer Claude Grahame-White also began preparations for an ocean flight.

But another project originating in the United States seemed to have the best chance. Backed by the philanthropist Rodman Wanamaker, a twin-engine Curtiss flying boat was designed for the Atlantic crossing. With a wing span of 74 feet, this was the largest craft yet constructed in the United States. Lieutenant John Henry Towers of the U.S. Navy, holder of the world nonstop distance and duration records for hydroaeroplanes—392 miles in 6 hours 10 minutes 38 seconds—was designated one of the pilots. He was to be accompanied by Lieutenant John Cyril Porte, an early British pilot of the Deperdussin. Porte allotted 20 hours in his calculations for the crossing. Wanamaker hoped that with one American and one British naval aviator sharing the controls, the flying boat America (described in the press as a high-powered craft, navigable in a strong wind though unable to rise from rough water) would give evidence of the friendship existing between the two countries and, incidentally, contribute to preserving world peace. In February 1914 the American benefactor declared:

Once across the ocean in one flight of an aircraft and the world will awake to a realization of the tremendous importance that aeronautics may prove to be to every nation. This year we are celebrating one hundred years of peace between Great Britain and the United States and it would be a fitting climax of the celebration if the two countries could link themselves by this international flight . . . demonstrating to the world that the time for disarmament of the nations is at hand, if for no other reason than because aeronautics has reached a stage where even the greatest dreadnaught battleships may become futile in their power.

Visit of British aviator Claude Grahame-White (in knickers and bow tie) to U.S. State, War, and Navy departments, October 14, 1910. His biplane, which he had landed on the street, is in the background. (National Archives)

While Wanamaker's prediction of the importance of planes in naval war-fare proved to be right, he was overly optimistic, as events soon showed, in regard to the contribution that aviation might make to world disarmament.

The advent of World War I finally put a stop to all private trans-oceanic projects on either side of the Atlantic. It was not until seven months after the Armistice that the £10,000 prize was won. On the afternoon of June 14, 1919, two Englishmen—Captain John Alcock and his navigator, Lieutenant Arthur Whitten Brown—took off from St. Johns, Newfoundland, in a twin-engined Vickers-Vimy bomber. Beset by rain, hail, snow, and icing on the wings most of the way, they barely made it to Ireland. The plane landed nose down in a bog near the little town of Clifden. For first piloting a flying machine nonstop across the Atlantic, the two aviators were knighted by King George V.

Alcock and Brown, in addition to making the first nonstop flight across the Atlantic, missed only narrowly the honor of being the first to pilot an aircraft of any type across the ocean. Just two weeks before their flight, Curtiss's Model NC-4 flying boat had completed a 24-day flight from Rockaway, Long Island, to Plymouth, England. The NC series of planes had been developed jointly by Curtiss and the Navy, and the transoceanic feat planned during the war; but the Armistice was signed before it could be attempted.

Then Admiral David W. Taylor said, "Though no longer a fight, there will still be a flight." By May of 1919, Models NC-1, NC-2, NC-3, and NC-4 were poised at the Rockaway Navy Base. An accidental fire during fueling of the NC-2 kept that craft out of the air; but the others took off on May 8. On the preliminary hop to Halifax the NC-4 developed engine trouble and put in at Chatham, Massachusetts, for repairs. However, it caught up with the others at Trepassey Bay, Newfoundland, in time to join them in the takeoff for the Atlantic flight on the 16th. En route to the Azores, the NC-1 and NC-3 alighted on the water to take their bearings; but damage suffered in landing prevented their resuming flight. The crew of the NC-1 were rescued by the ship Ionia before their plane sank, while the crew of the NC-3 safely reached Ponta Delgada in the Azores by sail-ing and taxiing. Meanwhile the NC-4, alighting first at Horta, winged on to Ponta Delgada. From there she flew to Lisbon, thereby completing the crossing of the ocean, and thence proceeded up the coast to Plymouth—arriving May 31. At the English city her crew joined with those of the NC-1 and NC-3 to receive acclaim for two heroic failures and the triumphant success of the NC-4—the first aircraft of any type to cross any ocean.

Eight years later, on May 21, 1927, Charles A. Lindbergh was to com-plete the first solo nonstop transatlantic flight—between New York and Paris. After his triumphal reception at Le Bourget, he declared: "I look forward to the day when transatlantic flying will be a regular thing. It is a question largely of money. If people can be found willing to spend enough money to make proper preparations, there is no reason why it can't be made very practical."

Europe now had something else to think of besides heroic overwater expeditions. The journalist Albert de Mun, analyzing the arms race in the *Echo de Paris,* had predicted early in 1913 that war was inevitable—that France and Germany were moving toward a clash "with the implacable certainty of destiny itself." An incident in the Coupe Internationale Aéronautique of October 1913 reflected the prevailing aura of suspicion. A balloon flying the German flag descended close to French soil at Chateaudun in order to ask directions. A group of peasants seized the balloon's guide rope and held it while a messenger was hastily dispatched to summon the gendarmes. When the police arrived, they found the balloonists' papers in order and allowed them to proceed—but the aeronauts were out of the race. According to the regulations, touching of the guide rope by persons other than the pilot or his aide constituted a technical landing and disqualified the competitor.

Yet most people shrank from the thought of war. The aeroplane that carried Garros across the Mediterranean had been hailed as a "messenger of peace." It was hard to look at the flying machine wholly from a military viewpoint; and few ventured to predict what role aviation might play in the event of conflict. Had not the commandant of the French Staff College decided in 1910 that "the aeroplane may be all very well for sport, but for the Army it is useless"? The attitude of the professional soldiers was in accord with international convention. In 1899, the Hague Peace Conference had gone on record to prohibit the discharge of projectiles or explosives from the air in time of war. While the delegates were thinking in terms of balloons and dirigibles, they were farsighted enough to include all aircraft, present and future. By international law, air vehicles were to be limited to a passive role, such as reconnaissance.

The first French air squadron was formed only reluctantly after the military maneuvers of 1911. As the historian Pierre Chastenet observed, military men were generally distrustful of the new invention; with no armament, no wireless, no really reliable motor, it was not to be counted on for practical use. And if aviation received a minimum of support from the army, it got hardly any from the civil government. Encouragement came instead from private circles concerned with the safety of the nation—as, for example, when the newspaper *Le Temps* opened a public subscription in 1910 to supply the army with flying craft.

The British War Office, too, was skeptical, and had rebuffed sagacious men like Moore-Brabazon who had urged greater expenditures for aeronautical development. One indignant critic summed up the prevailing attitude: "The apathy of the British authorities—one might almost say their hostility—seems incredible. Every topic under the sun is discussed in Par-

Two French military planes: the Bréguet biplane (at right) and the Hanriot monoplane (far right).

liament, from the making of cider to coal strikes, but hardly a word is ever spoken about the aeroplane." A. V. Roe, convinced that military aircraft were to be indispensable, was building biplanes as early as 1912 with his brother Humphrey for possible military use; but they, too, found the War Office unsympathetic. Acidly the department stated that it "saw no possibility of using aeroplanes for war purposes" and so could offer the brothers no support.

But while England continued in seeming oblivion of her exposed position, France was soon forced to open her eyes to reality. By the summer of 1914, the world had come to accept as a matter of fact the French lead in aircraft construction, in number of licensed aviators, and in establishment of flight records. However, as the shattering events of that first week in August drew closer, German flyers and machines began to attract wide attention, showing that they were on a par with the best France had to offer.

As in the United States, the German aviation service was originally part of the signal corps. Fettered by traditional cavalry concepts, it developed only slowly and with many technical difficulties. In 1912, however, Captain F. H. Geerdtz, an army pilot who had been named commandant, began to agitate for the establishment of a separate air force. Geerdtz's idea was to ring the German border with aerodromes and hangars, like forts. The impetus to German aviation created by this plan became noticeable two years later. So air minded had Germany become by 1914 that it seemed to have confirmed the visionary pronouncement of its nineteenth-century novelist Jean Paul Richter: "Providence has given to the French the empire of the land; to the English that of the sea; to the Germans that of—the air!"

Although Germany had always been partial to monoplanes, such as the Etrich or Rumpler Taube, biplanes specifically adapted to military use now began to dominate the skies. World records for altitude and duration

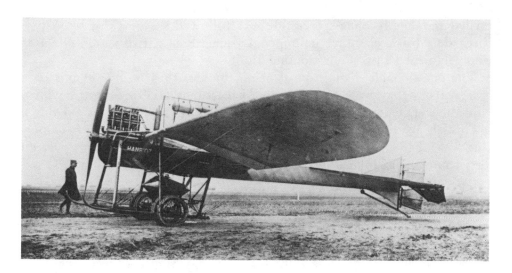

passed into German hands and were regularly raised. In the spring and early summer of 1914, Hellmuth Hirth as well as Ritter and Edler von Loessel, flying Albatros military models, pushed the altitude marks in various passenger-carrying categories to above 5000 meters. Then, on July 9, Gino Linnekogel, piloting a Rumpler biplane powered by a 100-hp Mercedes motor, ascended at Johannisthal to the rarefied height of 6600 meters, breaking the record of 6120 meters set on December 29, 1913, at St.-Raphaël, France, by Georges Legagneux in a Nieuport with a 60-hp Le Rhône motor. Linnekogel's record stood for only five days. At Leipzig on July 14, Heinrich Oelrich, in a new D.F.W. biplane (also Mercedes powered), rose to 7850 meters—more than a mile higher.

Meanwhile at Johannisthal, Werner Landmann captured the world record for distance, flying his Albatros (again with a 100-hp Mercedes motor) 1900 km on June 28, 1914. At the same time he smashed all previous records for duration by staying aloft 21 hours 50 minutes in a closed circuit. Almost unnoticed in the commotion of the approaching war, this mark was bettered two weeks later by another German, Reinhold Boehm, who kept an Albatros flying for 24 hours 12 minutes on July 10 and 11—the first plane in history to stay up for a full day and night.

Aeroplanes received their baptism of fire in the Italian campaign against Turkey during 1911–12. In support of Italy's expeditionary forces landing at Tripoli, Libya, an "aeroplane flotilla" was sent by sea under the command of Captain Carlo Piazza—winner of a gold medal in the Circuit of Italy held in September 1911. The initial unit consisted of two Blériots, two Nieuports, two Henry Farmans, and two Etrichs, together with ten officers, twenty-nine troopers, tent hangars, and a quantity of materiel. After a series of trials Piazza flew over the enemy lines for an hour in a single-seater Blériot on October 23, 1911, in the first officially recorded wartime reconnaissance. He was followed the next day by Captain Riccardo Moiza (another gold-medal winner), in a two-seater Nieuport, who performed a two-hour scouting flight.

The questionable distinction of being the first to drop live bombs on an enemy from the air went to Second Lieutenant Giulio Gavotti, who on November 1 deposited a small picric acid charge weighing 4.4 pounds on a Turkish position at Ain Zara, and three more on the oasis of Tagiura. On November 5, 1911, Italian headquarters in Tripoli issued the first military communiqué to mention the aeroplane:

Yesterday Captains Moizo, Piazza, and [Leopoldo] De Rada carried out an airplane reconnaissance, De Rada successfully trying out a new Farman military biplane. Moizo, after locating the position of the enemy's battery, flew over Ain Zara and dropped two bombs into the Arab encampment. He found the enemy were much diminished in numbers since he saw them last time. Piazza dropped two bombs on the enemy with effect. The object of the reconnaissance was to discover the headquarters of the Arab and Turkish troops.

It was hardly a surprise that the enemy were "diminished in numbers" after realizing the hostile intent of the Italian flying machines. No aircraft were at the disposal of the Turks; but rifle fire was sometimes effective. On February 1, 1912, Captain Montú was reported wounded in flight at an altitude of about 1800 feet by guns from an Arab encampment at Tobruk; on March 13, Lieutenant Cannoniere was likewise hurt while flying. When a second aerial company was set up at Benghazi, it achieved two other firsts: Captain Alberto Marenghi Marengoon, on May 2, 1912, made the first nighttime reconnaissance flight; and the first night bombing attack in history was made on June 11 against a Turkish encampment.

French operations aimed at the subjugation of Morocco embraced the use of aeroplanes for scouting during the same period. In the Balkan War of 1912–13, Bulgaria hastily organized an aviation corps. This force proved effective on several occasions; the equipment included six Bristol monoplanes, one 70-hp Blériot XII, two Blériot XI's captured from the Turks, and some half-dozen miscellaneous machines temporarily hired, with their pilots, from other countries. The Rumanian army boasted several 80-hp Bristol monoplanes and Henry Farman biplanes, as well as two Blériots, a Nieuport, a Morane, and two Vlaicu machines. Serbia had seven military aeroplanes by the end of March 1913, while Turkey— on the opposing side—had a fleet of about a dozen R.E.P. and Harlan monoplanes and a couple of biplanes. Only one of these, however, seems actually to have been used.

On the North American continent, the earliest reported use of a plane for military purposes was by Charles K. Hamilton during the Moisant tour. In 1911, deviating from his circus routine, Hamilton made an observation flight from El Paso, Texas, over Ciudad Juárez, Mexico, while fighting was in progress between Mexican government troops and rebels. Late in 1913, Phil Radar, flying for General Huerta, and Dean Ivan

Lamb, flying for General Carranza, exchanged a dozen or so pistol shots over Neco, Mexico—the first recorded aerial dogfight in history.

On March 3, 1911, Congress made its first appropriation for military aviation—$125,000 for the fiscal year 1912—and the Signal Corps immediately ordered five new aeroplanes. The first aerial squadron in the United States was organized on March 5, 1913, at Texas City, Texas—a site selected because of the continuing tense border situation resulting from the revolution in Mexico. The squadron included nine planes, made by Wright, Curtiss, and Burgess. On March 31, Lieutenant William C. Sherman produced the first aerial map made in a military aircraft. On an eighteen-foot roll of paper he recorded a great amount of graphic intelligence—representing railroads, wagon roads, streams, woods, and prairies—a remarkable achievement in a day when almost any flight was news.

The United States, of course, was not to enter the European conflict until 1917; and American military authorities evinced the same lack of conviction as to the aeroplane's worth in war that their European counterparts had exhibited before 1914. Various experiments were conducted at North Island, California (after the Signal Corps was moved from Texas City) as to the possibilities of airborne wireless, flares, and automatic pilots; but results were of little practical value. Tests were also performed using the Scott bombsight and bomb-release mechanism,

Two U.S. Army pilots, Lieutenants Milling (left) and Sherman (right), in a Burgess-Wright military tractor biplane. On March 28, 1913, during a flight between Texas City and San Antonio, Texas, Sherman produced the first aerial map made from a military aircraft. (National Archives)

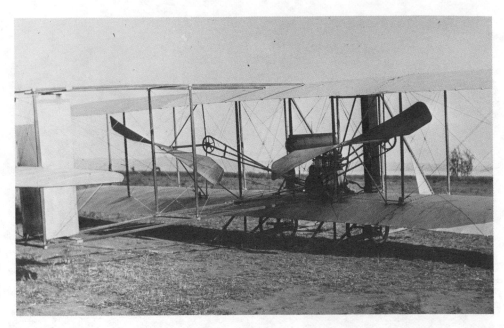

An Air Force Wright Model B biplane at San Diego, California, 1913. (U.S. Air Force)

results of which made a deep impression on the "young flyers" at North Island. However, they were not so impressive to Army officers of higher rank and lesser vision. When a machine gun was fired from a plane for the first time, by Captain Charles de Forest Chandler at College Park, Maryland, on June 7, 1912, one staff officer made it clear that aeroplanes were "suitable only for reconnaissance" and that thoughts of air battles were "purely the product of the young flyers' fertile imaginations."

The United States trailed far behind other nations in acquiring or developing military aircraft. Although the famed Curtiss Model JN-4 biplane, a tractor (nicknamed the "Jenny"), made its first flight in 1914, American air power when hostilities broke out abroad was roughly comparable to that of little Switzerland.

When the Great War began, the air strength of the different nations showed a wide discrepancy. Military historians disagree as to the precise statistics; but according to Arthur Sweetser, in a compilation published in 1919, the number of military aeroplanes available in the fiscal year 1914 was as follows:

Nation	Planes
France	260
Russia	100
Germany	46
England	29
Italy	26
Japan	14
United States	6

On both sides aerial squadrons were built up of a heterogeneous collection of planes and engines. Most of the machines had not been constructed for specialized military functions and, except for scouting purposes, were not especially useful at first. Few if any of the planes on either side could do better than 60 m/hr. France depended on the Nieuport, Deperdussin, Morane, R.E.P., and Blériot monoplanes, along with Maurice Farmans, Voisins, Caudrons, and Bréguets among the biplanes. Germany had taken to the air with Rumpler Taubes, Gothas, German-made Bristols, and Fokker monoplanes, as well as the Albatros, L.V.G., Aviatik, and Otto biplanes. The Russians had acquired aircraft from more than twenty different manufacturers; these were principally Henry Farmans and Nieuports, but also included Curtiss water-based machines from the United States. During the first few weeks, under the exigencies of wartime flight, disastrous crashes were common. Makeshift bases hurriedly prepared were responsible for a high rate of landing and takeoff accidents. Pilots had little or no experience in cross-country navigation and usually lacked adequate maps; if they became lost and ran out of fuel, as often as not they would fall behind enemy lines.

The British, as a first effort, sent sixty-four aeroplanes to France: the

A Russian squadron with the Sikorsky Ilia Mourometz and a 1000-pound bomb at the Eastern Front during World War I. (I. I. Sikorsky)

B.E. (Blériot Experimental), designed by Geoffrey de Havilland, with a 70-hp Renault engine—later produced by the thousands; the Avro Model 504 biplane, with an 80-hp Gnôme motor, first built by A. V. Roe in 1913 and highly regarded as a trainer; the Sopwith Tabloid scout (ancestor of a long line of wartime Sopwiths), likewise powered by a Gnôme; and a handful of Henry Farmans and conventional Blériots. Ten planes were lost and at least five pilots killed in this initial transport operation. The small Belgian force of two dozen French planes was almost wiped out before the fighting bogged down into trench warfare. Thereafter the quest began for a combat design that emphasized high rate of climb, maneuverability, and an unobstructed view from the cockpit.

At first, an air of chivalry prevailed among flyers at the front. Many pilots had been recent competitors in international events, and those on one side were frequently recognized by and well known to those on the other. Soon, however, it became imperative to prevent an observation plane from returning with its reports; and the era of the dogfight began—the opponents pitted against each other like medieval knights. Revolvers became part of every airman's equipment; rifles, shotguns, and machine guns followed. One device was to dangle a brick from the end of a rope to break an opponent's propeller. On October 5, 1914, a French air mechanic named Quenault, flying as observer in a Voisin biplane piloted by Joseph Frantz, downed a German Aviatik, in the first aerial combat resulting in an Allied success. As proof of victory Frantz brought back the helmet and identification papers of the enemy, receiving from his superiors in return a bouquet of flowers, a pheasant, and the Cross of the Legion of Honor.

It was the custom on both sides, when a noted flyer went down, for his opponents to render appropriate honors by dropping a wreath. After Adolphe Pégoud had been brought to earth, a bullet in his neck, during an encounter between his Nieuport and a large, new two-seater Taube on September 1, 1915, the tribute read: "To The Aviator Pégoud, Fallen In Combat For His Country. Homage Of The Victors." On the Eastern Front, German pursuit planes destroyed the Ilia Mourometz XVI with all on board on September 25, 1916, when it ventured too far over enemy lines. A few days later a note was dropped by the Germans, stating that the four-man crew had been buried with military honors and giving the location of the tomb, on which a cross of Russian Orthodox style had been placed.

Reciprocal courtesies of this kind were to stud the progress of the war. Heroes of the air were immortalized in stories told about the Lafayette Escadrille, a corps of American aviator volunteers; and by awards of the French Legion of Honor and of the German order Pour le Mérite—the latter nicknamed the "Blue Max," after Max Immelmann, who made famous the half-loop, half-roll "Immelmann turn." It is paradoxical that Germany's highest recognition for military valor was instituted by Frederick the Great, whose fondness for things French was responsible for the designation "Pour le Mérite." Fourteen flyers won this coveted medal.

Goaded by the insistent demands of combat pilots, constructors were under constant pressure to improve the capabilities of their machines. The Voisin, armed by a machine gun in front of the cockpit fired by the observer, evolved into a heavy bomber—as did the Caudron; and the Maurice Farman was used extensively for scouting. But these were clumsy machines to handle in tight corners. The *cage à poules* was a relatively easy prey for the fast Fokkers that gradually supplanted the Taubes, so heavily relied on at the start.

Even the Morane-Saulnier was outpaced by the Fokker; but the latter in turn met a match in the Nieuport—powered by a reliable motor which, in the view of many pilots, performed better than the Gnôme. This was the 110-hp Le Rhône—a wartime version of the air-cooled, seven-cylinder, rotary type that had lifted so many planes off the ground in the past. The new Le Rhône, less extravagant of oil and gas, was a better foil to the stationary, water-cooled engines (such as the Mercedes) universally in use on the other side.

But monoplanes were too dangerous structurally for the requirements of combat flying. Eighteen months after the war had begun, almost all flight units on both sides consisted principally of biplanes. The single wing did not reappear until near the end of the war—and then in a greatly reinforced version.

The German answer to the Nieuport was the glittering red Fokker triplane—made famous by Baron Manfred von Richtofen—which, although it sacrificed speed for climbing power and maneuverability, was almost invincible in the proper hands. Later, to bolster French resources, came the Spad. Faster than anything yet seen (though not as quick as the triplanes on turns or in climbs), this successor to the Blériot and Deperdussin bore the brunt of saving France—much as the Hurricanes and Spitfires of a later day were to save Britain.

The era of the early flying machines ended with the opening campaigns of World War I. The war stepped up the development of aeroplanes and engines to an incalculable degree. It opened the eyes of the world to the limitless possibilities of air travel, turning the "toy" of the sportsman and the exhibition flyer into a practical vehicle and preparing the way for private flight on a scale unimagined before hostilities began.

Unhappily, it also turned the flying machine into an instrument of terror and destruction, as an indispensable adjunct to the armed forces of every nation—an incarnation that many of its originators foresaw but hoped would never happen. The aeroplane made possible warfare in three rather than two dimensions, and thus revolutionized completely all earlier ideas of strategy and tactics.

The planes that emerged at the end of the conflict in 1918 were very different from the primitive contrivances that had entertained the crowds at aviation meets, put through their restricted paces by the resolute, brave—even fanatical—vanguard of early birds. For when peace came, the forerunners of modern commercial airliners were ready to take to the air; and the prospect for the future of flying was fundamentally changed.

At a banquet sponsored by the Aero Club of America in New York in 1916, honoring Alberto Santos-Dumont, the guests were handed sealed envelopes bearing the inscription: "Please do not open until Mr. Santos-Dumont has been introduced." When the great pioneer rose to acknowledge the ovation that greeted him, they found the following message from Henry A. Wise Wood, the club's president:

A few years ago they smiled at the dirigible, and it is shaking the World's imperial city; they laughed at the aeroplane, and it is directing the armies of Europe. They deride the overseas aerial liner, but I declare to you, my friends—not as a dreamer speaking in rhapsody but as a cold and calculating engineer—that the day is at hand when the building of such a vessel may be begun, and is not far distant ere that liner shall, at established intervals of time, set forth for every capital. Knowing neither land nor sea, nor reef nor ice, nor coast nor boundary, these winged, swift, and ghostly argosies shall make of every town an aerial port, of every hamlet a haven of safety on the coast of that vast ocean of which its earliest navigator sits amongst you here. Then shall there be realized the dream that has led us on, and there be drawn into a single community possessed of a common consciousness the national families of all the Earth. Than this, if there be a vision bigger with human good, I know it not.

The progress in human flight had only just begun. Man's first step in freeing himself from the earth was his invention of the aeroplane; and the crude vehicles of those early days started him on flights which were soon, indeed, to surpass those of the birds. From his exploration of the atmosphere he advanced to the gaining of loftier goals: the short hops of the pioneer flyers became, in amazingly brief time, the prelude to travel in outer space. Today, the success of all voyages beyond the earth rests on the heroic contribution of those who showed the way.

APPENDIXES

First Aeroplane Flights in Different Countries

B = biplane   M = monoplane

Date	Country	Aviator	Machine	Place
1903				
Dec. 17	U.S.A.	Wright brothers	Wright (B)	Kitty Hawk, N.C.
1906				
Sept. 12	Denmark	J. C. Ellehammer	Ellehammer (B)	Lindholm
Sept. 23	France	A. Santos-Dumont	Santos-Dumont (B)	Paris
1908				
May 15	Italy	L. Delagrange	Voisin (B)	Rome
May 26	Belgium	H. Farman	Voisin (B)	Ghent
June 8	England	A. V. Roe	Avro (B)	Brooklands
June 18	Holland	E. Lefebvre	Wright (B)	The Hague
October	Scotland	L. Gibbs	Dunne (B)	Perthshire
Nov. 24	Germany	A. Zipfel	Voisin (B)	Berlin
1909				
Feb. 23	Canada	J. A. D. McCurdy	Silver Dart (B)	Baddeck, N.S.
Apr. 23	Austria	G. Legagneux	Voisin (B)	Vienna
July 25	Russia	Van der Schrouff	Voisin (B)	Odessa
July 29	Sweden	G. Legagneux	Voisin (B)	Stockholm
Oct. 17	Hungary	L. Blériot	Blériot (M)	Budapest
Oct. 30	Rumania	L. Blériot	Blériot (M)	Bucharest
Nov. 15	Algeria	R. Métrot	Voisin (B)	Blida
Dec. 2	Turkey	P. de Caters	Voisin (B)	Constantinople
Dec. 15	Egypt	P. de Caters	Voisin (B)	Abbassia
Dec. 28	South Africa	Kijmmerling	Voisin (B)	East London

Date	Country	Aviator	Machine	Place
Dec. 31	Ireland	Ferguson	Ferguson (M)	Hillsborough
December	Portugal	A. Zipfel	Voisin (B)	————
December	Mexico	A. Braniff	Voisin (B)	Mexico City
1910				
February	Australia	H. Houdini	Voisin (B)	Sydney
Feb. 2	Argentina	H. Brégi	Voisin (B)	Burzaco
February	Brazil	G. Ruggerone	H. Farman (B)	Havana
February	Cuba	H. Brégi	Voisin (B)	————
Feb. 10	Spain	J. Mamet	Blériot (M)	Barcelona
Mar. 13	Switzerland	Engelhard	Wright (B)	St.-Moritz
Apr. 21	Transvaal	Kijmmerling	Voisin (B)	————
Aug. 21	Chile	Cesar Copetta	Voisin (B)	Santiago
Oct. 14	Norway	C. Cederstrom	Blériot (M)	Etterstadt
Dec. 15	Indochina	C. Van den Born	H. Farman (B)	Saigon
Dec. 17	India	H. Pequet	Humber (M)	Allahabad
Dec. 19	Japan	Y. Tokugawa	H. Farman (B)	Tokyo
1911				
Feb. 9	New Zealand	V. Walsh	Howard-Wright (B)	Papakura
February	Peru	J. Bielovucic	Hanriot (M)	Lima
Feb. 21	China	R. Vallon	Sommer (B)	Shanghai
December	Madagascar	M. J. Raoult	Blériot (M)	Tananarive
————	Korea	J. Mars	Curtiss (B)	————
————	Philippines	J. Mars	Curtiss (B)	————

(Compilation from various sources, including *The Boys' Book of Aeroplanes,* by T. O'B. Hubbard and C. C. Turner, Frederick A. Stokes Company, New York, 1913.)

Aeroplane Fatalities and Their Causes

In the first three years during which fatalities were recorded—1908, 1909, and 1910—there were thirty-three crashes resulting in death to the pilot, in two of which a passenger was also killed. A thousand persons, more or less, learned to fly during this period. Thus the mortality rate of around three percent was not unduly high, compared with such hazardous sports as mountain climbing—in which ninety people were reported to have died in 1910 alone. Later, as more people took up flying, accidents increased proportionately. Death in the sky came to afford an all-too-frequent excuse for sensational headlines.

Of the first thirty-three fatal crashes, thirteen were due to structural weakness in the machine, and another three to failure of the control mechanism. Among the former, for want of a better explanation, may be included the disappearance of Cecil Grace. Eight crashes could be ascribed to the pilot's loss of control; three were caused by sudden gusts of wind; and three, ironically, were the result not of failure in flight but of mishaps on the ground. Only one was clearly caused by motor trouble—a common cause of catastrophe in later years. Two are believed to have been brought on by physical weakness or illness of the pilot while flying.

By the end of 1910, flyers of eleven nations had incurred fatal crashes; these included 10 Frenchmen, 6 Americans, 5 Italians, 4 Germans, 2 Englishmen, 2 Belgians, 2 Spaniards, 1 Russian, 1 Dutchman, 1 Hungarian, and 1 Peruvian. The tally bore grim testimony to the current widespread interest in flying. A year later the number killed in crashes had jumped to 76; on October 1, 1912, the total stood at 191. Pilots followed one another to the grave in dismaying succession. Anthony Fokker was to note, "Every flying field I have known is soaked with the blood of my friends and brother pilots. . . . My memory is one long obituary list."

The first man to be killed in a plane crash was Lieutenant Thomas E. Selfridge, a passenger in a biplane being tested by Orville Wright at Fort Myer, Virginia, on Sept. 17, 1908. For a year after Selfridge's death there were no further fatal accidents. Then the public was stunned by the deaths of four pioneers in rapid succession. Each had been on the program at Rheims, and three of the four had thrilled the crowds with their exceptional daring and skill. Their demise made it clear that man would not realize his dream of soaring through the clouds without paying a high price.

The first of this group to die was Eugène Lefebvre, whose exuberant evolutions in the air at Rheims had won him fame as the world's first stunt flyer. On the morning of September 7, 1909, shortly after the close of the Rheims meet, the twenty-seven-year-old Lefebvre decided to test a Wright machine—the first to be fitted with wheels attached to the skid undercarriage. The takeoff from the aerodrome at Juvisy went smoothly. But the machine had flown only a few minutes, at relatively low altitude, when it went into a dive and struck the earth with such force that Lefebvre perished instantly. It was difficult to decide what caused the crash, but most witnesses agreed that the controls had jammed.

Two weeks later, on September 22, Captain Ferdinand Ferber, of the Voisin team, lost his life at Boulogne-sur-mer in a freak accident. He had finished a flight and was rolling along rough terrain when the wheels of his biplane plunged into a ditch. The shock dislodged the engine from its bed, and it fell directly on the aviator. Ferber at first appeared only shaken, able to speak and move; but the internal injuries he had incurred were so severe that he succumbed shortly afterward.

On December 6, Antonio Fernandez, who had built a light biplane along Curtiss lines with some variations of his own, died in a crash at Nice. A Spaniard who resided in Paris, Fernandez had entered the Rheims meet but was not able to get his machine ready in time to participate. Afterward he made a number of successful tests on the Riviera. When he ascended for the last time, watched by his young wife, the machine broke up at an altitude of several hundred feet—and Fernandez fell to his death. Witnesses could not agree on the cause; an exploding engine cylinder may have been responsible.

Less than a month later, on January 4, 1910, Léon Delagrange became the fourth Rheims veteran to die. Flying a Blériot monoplane at Bordeaux, he had begun a series of experiments to determine whether the 50-hp Gnôme engine was better than the 25-hp Anzani that Blériot himself had used when he crossed the Channel. The day before, on January 3, using the Anzani, Delagrange had inaugurated the new aerodrome of Croix d'Hins at Bordeaux with an 8-minute flight, despite a ground haze so thick that spectators had difficulty finding him after he landed. On the 4th, substituting the Gnôme, he took off in somewhat gusty weather; the higher-powered and therefore faster machine suffered strains not well understood at the time. A wing buckled, and the plane fell from a height of 60 feet, killing Delagrange instantly.

The first double death involved two Italians: Lieutenant Enrico Cammarota (a military engineer) and an army private named Castellani. Both died in the wreck of their Henry Farman biplane when it cracked up while making a sharp turn at the Centocelle aerodrome on December 3, 1910. The first fatal crash involving two different aeroplanes was recorded on June 19, 1912. On that date Captain Dubois and Lieutenant Peignan of

Aeroplane Fatalities, 1908-1910

Date	Aviator	Nationality	Machine	Place	Cause
1908					
Sept. 17	Lt. Thomas E. Selfridge	American	Wright biplane	Ft. Myer, Va.	Breakage of propeller
1909					
Sept. 7	Eugène Lefebvre	French	Wright biplane	Juvisy	Failure of control mechanism
Sept. 22	Capt. Ferdinand Ferber	French	Voisin biplane	Boulogne-sur-mer	Crackup on ground
Dec. 6	Antonio Fernandez	Spanish	Fernandez biplane	Nice	Breakage of machine in air
1910					
Jan. 2	Aindan de Zoseley	Hungarian	de Zoseley	Budapest	Loss of control
Jan. 4	Léon Delagrange	French	Blériot monoplane	Bordeaux	Wing collapse
Apr. 2	H. Leblon	French	Blériot monoplane	San Sebastián	Motor failure
May 13	Hauvette-Michelin	French	Antoinette monoplane	Lyons	Collision on ground with pylon
June 18	Thaddeus Robl	German	Aviatik biplane	Stettin	Wind gust
July 3	Charles Louis Wachter	French	Antoinette monoplane	Rheims	Wing collapse
July 12	C. S. Rolls	English	Wright biplane	Bournemouth	Breakage of machine
July 13	Daniel Kinet	Belgian	H. Farman biplane	Ghent	Broken rudder
Aug. 3	Nicholas Kinet	Belgian	H. Farman biplane	Stockel	Rear wire tangled in motor
Aug. 20	Lt. Pasqua Vivaldi	Italian	M. Farman biplane	Rome	Loss of control
Aug. 27	C. van Maasdyck	Dutch	Sommer biplane	Arnhem	Loss of control

the French army met death near Douai when their machines collided at an altitude of about sixty meters and fell locked together to the ground.

Of the first thirty-three fatal accidents, the twin-propellered Wright biplane accounted for almost a quarter. As a consequence of this unlucky record, the American machine came to be looked upon by many flyers as a "killer." In addition to Ralph Johnstone and Arch Hoxsey (the "Heavenly Twins"), who died in 1910, pilots killed in Wrights over the next few years included such skilled Americans as Calbraith Rodgers, Howard Gill, A. L. Welch, Phil O. Parmelee, and Max Lillie—holder of Expert Aviator's Certificate No. 1. In nearly every instance the aviator, sitting

Date	Aviator	Nation-ality	Machine	Place	Cause
Sept. 25	Edmond Poillet	French	Savary biplane	Chartres	Wind gust
Sept. 27	Georges Chavez	Peruvian	Blériot monoplane	Domodossola	Numbness due to cold
Sept. 28	—— Plochmann	German	H. Farman biplane	Hausheim	Loss of control
Oct. 1	—— Haas	German	Wright biplane	Wellen	Breakage of propeller chain
Oct. 7	Capt. Mazievitch	Russian	H. Farman biplane	St. Petersburg	Failure of control mechanism
Oct. 23	Capt. Madiot	French	Bréguet biplane	Douai	Illness in flight
Oct. 25	Lt. Mente	German	Wright biplane	Magdeburg	Loss of control
Oct. 26	Fernand Blanchard	French	Blériot monoplane	Paris	Failure of control mechanism
Oct. 27	Lt. Saglietti	Italian	Italian Wright biplane	Rome	Loss of control
Nov. 17	Ralph Johnstone	American	Wright biplane	Denver, Colo.	Breakage of machine in air
Dec. 3	Walter Archer	American	Archer	Salida, Colo.	Loss of control
Dec. 3	{ Lt. Cammarota / Pvt. Castellani	Italian	H. Farman biplane	Rome	Breakage of machine
Dec. 22	Cecil Grace	English	Short biplane	English Channel	Structural failure
Dec. 28	Guilio Picollo	Italian	Antoinette monoplane	São Paulo	Blow from tail of machine on ground
Dec. 28	{ Aléxandre Laffont / Marquis de Pola	French / Spanish	Antoinette monoplane	Paris	Wing collapse
Dec. 30	Lt. de Caumont	French	Nieuport monoplane	St. Cyr	Loss of control
Dec. 31	John B. Moisant	American	Blériot monoplane	New Orleans, La.	Wind gust
Dec. 31	Arch Hoxsey	American	Wright biplane	Los Angeles, Cal.	Breakage of machine in air

(From "Chronology of Aviation," prepared for *The World Almanac* by Hudson Maxim and William J. Hammer, New York, 1911.)

on the leading edge of the plane with no protection forward, was pitched out when the plane crashed. Fatal injuries were incurred in this way also by pilots of the Henry Farman, the Sommer, and their imitators. Maurice Farman and Voisin sought to minimize the danger (as well as reduce head resistance) by enclosing the pilot in a rudimentary nacelle; and Maurice added long, curved skids at the front of his machine to cushion the shock. The pusher planes also were dangerous because of the location of the engine at the back. In a crash the engine was likely to tear loose and plunge forward, killing or maiming the pilot.

Monoplanes, although safer in some respects (most were tractors, with an engine and a certain amount of woodwork between the pilot and the ground), presented hazards of their own. The wings of the Antoinettes and the Blériots frequently broke off as a result of inadequate bracing. And the pilot, instead of being thrown out on impact, might be telescoped into the front of the machine, breaking one or both legs.

Many of the injuries and deaths in the early years might have been prevented if pilots had used seat belts. Incomprehensible as it may seem, it required numerous object lessons to drive home the fact that an aviator and his machine could be soon parted unless securely strapped together. Among the few machines equipped with this elementary safety feature was the R.E.P. The "racing shell" fuselage of the Antoinette, too, had a belt to hold the aviator in his perch; but it was a rigid belt, without elastic and riveted to the hull, so that it tended either to break as a result of impact or else hold fast and inflict internal injuries on the occupant.

The eight accidents, out of the fatal thirty-three, ascribed to "loss of control" may also have resulted from the pilot's lack of experience. For instance, it was not always realized that to pull out of a nose dive, an even steeper dive was necessary. People thought it was "dangerous to fly too high"—whereas altitude actually made for safety if a plane started spinning. One must also remember that the usual object of designers was lightness of construction at any cost; that the air was not then fully understood as a medium of travel; and that the art of maintaining stability had first to be learned, as in bicycle riding. In the Henry Farman and Sommer types of machine, for example, the propeller directed air currents against the tail, causing it to lift; if the motor stopped and a gliding angle was not immediately assumed, the tail would drop and the aeroplane would quickly become unmanageable. In general, the old machines had one inherent advantage: their landing speed was so slow that a crash could be sustained without necessarily fatal results. The infrequency of fire in a wrecked aircraft was one illustration of this fact.

A general survey made in 1913 indicated that approximately 40 percent of the accidents then occurring were due to breakage in mid-air; 28 percent to loss of control; 20 percent to wind gusts; 12 percent to mechanical failure of the control system; and 10 percent to motor failure or to mishaps on the ground. A generation later, "loss of control" would probably have been called "pilot's error"; and structural weakness would account for only 10 to 15 percent of the crashes. This record does not compare very favorably with that of the early days of space travel—an infinitely more risky form of navigation, it would seem. In 1900 hours logged by United States astronauts in space up to November 15, 1966, not a single man had received so much as a scratch. Undoubtedly the laborious process of getting man off the ground and above the clouds during the early years of flight contributed enormously to the studied precision with which he now orbits the earth. In that light, the sacrifices of the pioneers were not in vain.

The Pioneers:
Their Later Lives

The war claimed the lives of many pioneers who had naturally gone into that branch of the armed forces where they could serve their countries best. One or two of them compiled adventurous accounts for themselves before the end came. Roland Garros, shortly after hostilities opened, invented a method of shooting a machine gun through the blades of a propeller, confounding the enemy with this unconventional use of firepower. When he and his plane were captured on April 19, 1915, after bombing a German train near Courtrai, the secret weapon was revealed to the other side. On February 14, 1918, disguised as a German officer, he escaped with Anselme Marchal (the only Allied pilot who had flown over Berlin) and fled to Holland. Garros quickly returned to the front, a chevalier of the Legion of Honor. He lost his life in the wake of a dogfight with a group of Fokkers on October 5, 1918—barely a month before the Armistice. The exact cause of his death is still a mystery, though the theory has been advanced that a malfunction of his machine gun shattered the propeller and fatally damaged the plane. Garros was buried in the cemetery at Vouziers; nearby, a modest shaft erected by his father marks the exact place where his Spad crashed. A sports stadium dedicated to his name is a well-patronized feature of the Parisian scene today.

Marc Pourpe was one of the first to die, when his machine crashed on December 2, 1914, during a reconnaissance in stormy weather. Brindejonc des Moulinais, who distinguished himself in the First Battle of the Marne, fell from the clouds in an unexplained crash on August 19, 1916, and was killed. Eugène Gilbert early became an ace. Interned for a time in Switzerland after a forced landing in that neutral nation, he escaped on his third try. Thereafter, however, he was prohibited by ear trouble from flying at high altitudes and was put in charge of a group that tested and took delivery of new military machines. On May 16, 1918, he fell to his death in a prosaic aerodrome accident—brought down by a broken stabilizer not far from the spot where he had taken off.

Jules Védrines survived the war, serving his country well by landing French agents behind the German lines and picking them up again. His vivid career came to an end in 1919, when he was killed in a crash while trying to demonstrate the commercial possibilities of regularly scheduled Paris–Rome flight.

Jacques de Lesseps, one of the heroes at Belmont Park, also came through the war unscathed, and possessor of the Croix de Guerre and seven citations—only to be lost in a fog with his mechanic over Quebec on October 18, 1927. His body, recovered off Newfoundland nearly two months later, is buried in the cemetery at Gaspé—not far from a monument erected to his memory by Canadian friends.

Robert Loraine, the actor-aviator, also survived the war. Serving as a pilot-observer, he reached the rank of lieutenant colonel. After peace was declared, Loraine became a broadcaster. He died in 1935, his health undermined by wounds received in combat.

The record shows that most pioneer aviators met an early death as a result of crashes in which they were involved. An exception was Hubert Latham, whose brilliant flight achievements were terminated by a totally unrelated mishap on the ground. On July 16, 1912, while on a big-game hunt near Ft. Lamy in French Equatorial Africa, Latham was killed in the charge of a Congo buffalo he had shot but only wounded. Some flyers, of course, lived relatively longer, surviving crashes as if under a magic spell. And a few lived out a normal life span, retiring from active flying while still favored by fortune, leaving to others the task of carrying on experiments and setting new records.

Among this last group was Orville Wright, who outlived his brother by some thirty-six years. When he died in Dayton in 1948 at the age of almost seventy-seven, he had seen his simple ideas of flight more than fulfilled in giant airliners spanning seas and continents with routine ease. The Wrights have always shared the reverence of the world as coinventors of the aeroplane. Wilbur was elected in 1955, and Orville in 1965, to the Hall of Fame for Great Americans at New York University.

Glenn Curtiss spent his later years dissociated from aviation as a real-estate developer in Florida. The patent litigation in which he and the Wrights had been engaged, and which had so unhappily poisoned the atmosphere of early American aviation, left him an embittered and vindictive man. He died in 1930, at the age of fifty-two.

Of Curtiss's companions in the Aerial Experiment Association, Frederick W. Baldwin died at his home near Baddeck on August 7, 1948, aged sixty-six. John McCurdy continued in aviation for many years, but gave up flying in 1916 when his eyesight began to weaken. He was appointed lieutenant governor of Nova Scotia in 1947. McCurdy died in a Montreal hospital on June 25, 1961, at the age of seventy-four.

Other American pioneers died of causes unconnected with aviation. To mention only a few, Charles K. Hamilton—despite his prediction, "We shall all be killed if we stay in this business"—was a victim of pneumonia at Hartford on January 22, 1914; Earle Ovington, the original mail carrier, died after a lingering illness in Los Angeles on July 21, 1936, aged fifty-six; and Glenn Martin succumbed to a cerebral hemorrhage at his home near Baltimore on December 4, 1955—not quite seventy years of age. James C. Mars, whose careers after he gave up flying included airport construction, pilot training, gas engineering, and real estate, died of a heart ailment in the Los Angeles Veterans Hospital on July 25, 1944, aged sixty-eight. Harry N. Atwood was believed to be the last-surviving early pilot of the Wright biplane when he passed away at the age of 83 on July 14, 1967, at Murphy, N.C. Dr. William Whitney Christmas, prolific inventor with a credit of 300 patents, lived longer than most people can expect. Christmas was ninety-four years old when he died of pneumonia in New York's Bellevue Hospital on April 14, 1960. Yet he was outlived by Gustavus Green, builder of Britain's first successful airplane engine. Green, who had started out (like the Wrights) in a bicycle shop, in 1908 evolved the 60-hp engine which a year later powered the machine used by Moore-Brabazon for the first circular flight of 1 mile in Great Britain. Green died at his home near London on December 30, 1964, aged ninety-nine.

Some U.S. Army aviators survived to an advanced age. Brigadier General Frank P. Lahm, the second to solo in the Signal Corps's initial flying machine, died of a stroke at Sandusky, Ohio, on July 7, 1966—eighty-five years old. And Major General Benjamin D. Foulois, at the time of his death at Andrews Air Force Base near Washington, D.C., on April 26, 1967 (aged eighty-seven), was engaged in writing down his recollections as the country's oldest military pilot.

Claude Grahame-White, the British star of America's international meet at Belmont Park in 1910, died at Nice on August 19, 1959—three days before his eightieth birthday. He had been commissioned in the Royal Naval Air Service in 1914, and took part in the first air raid on German-held bases in Belgium. His postwar aviation activities made him a wealthy man, and he owned several homes in both England and the United States. The author of several volumes on aviation, Grahame-White was describing his flying experiences in an autobiography at the time of his death; it was completed for him by his friend Graham Wallace, in 1960. A. V. Roe, a contemporary of Grahame-White, also lived to the age of eighty, while Geoffrey de Havilland, died at his home in Watford, England, on May 21, 1965, at the age of eighty-two. De Havilland was knighted for his contributions to the development of the aeroplane in both war and peace.

Alberto Santos-Dumont died in his native Brazil on July 23, 1932—an unhappy man. Saddened by the use for destructive purposes of the machines he had helped to perfect—and which he had hoped would bring peace to mankind—he could not free his mind of the melancholy induced by the war years. He was further shaken by a tragic accident that occurred during a celebration arranged in his honor. On December 23, 1928, Santos-Dumont was returning home to South America on the liner Arcona; as the ship entered Rio de Janeiro's harbor, he watched the spectacular crash of a large plane, filled with Brazil's intellectual elite, which had flown out to the steamer in welcome to the nation's "father of aviation." In 1932, Santos-Dumont was again sickened in soul when the state of São Paulo rose in revolt against the rest of the country and employed the aeroplane for bombing attacks. This last was too much for the illustrious and sensitive pioneer; he took his own life. Today Santos-Dumont is a national hero—his memory kept fresh by the airport at Rio that bears his name.

Leaders of French aviation also lived to an advanced age. Louis Blériot died of heart failure in Paris on August 2, 1936, at sixty-four—while Alessandro Anzani, designer of the engine that powered Blériot's historic flight across the Channel, was seventy-nine when he died near Caen on July 24, 1956. Louis Bréguet was felled by a heart attack in Paris on May 4, 1955, aged seventy-five. Esnault-Pelterie died at Nice on December 6, 1957, over seventy-six years old. His name is perpetuated by the designation of the street on which the inventor lived as Rue Robert Esnault-Pelterie. Other French pioneers lived into their eighties. Henry Farman had reached the age of eighty-four when he succumbed, after a long illness, in Paris on July 17, 1958. But Henry was outlived by his brother Maurice, who also died in the French capital, on February 25, 1964. Colonel Jacques Balsan, one of the organizers of the Lafayette Escadrille in World War I, died at his home in New York in 1956 at the age of eighty-eight. Roger Sommer, too, died at eighty-eight; he passed away on April 15, 1965.

Equally long lived were some famous women flyers. Marie Marvingt, known as "The Fiancée of Danger," terminated a career filled with excitement and sport after eighty-eight years, on December 15, 1963. "Mademoiselle Marie," daughter of a postmaster, was a prize-winning bicyclist, mountain climber, skier, swimmer, boxer, wrestler, judo expert, war correspondent, writer, nurse, surgeon, balloonist, and aviatrix, to mention but a few of her activities. The Marvingt name was best known, however, in the field of heavier-than-air machines. On November 27, 1910, in a bid for the Coupe Fémina at Mourmelon, she made a world record for women by flying 45 km in 53 minutes. At the age of eighty she piloted a jet-engined helicopter. "I don't think any man would put up with me for long," she declared, explaining why she never married. "I'm more interested in mountain climbing than in washing dishes."

Hélène Dutrieu, the pioneer Henry Farman pilot, passed away on June 26, 1961, at the age of eighty-four. Among her accomplishments were the first flight of 1 hour by a woman (December 15, 1909), to win the 2000-franc Coupe Fémina with a distance of 60 km 700 meters; the first flight by a woman with passenger, at Mourmelon on April 20, 1910; and a world altitude record for woman flyer with passenger—1300 feet—achieved at Ostende in September of that same year.

Chart of Metric Equivalents

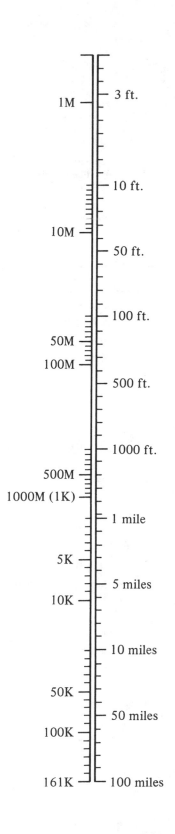

1M	3 ft.
	10 ft.
10M	50 ft.
	100 ft.
50M	
100M	500 ft.
	1000 ft.
500M	
1000M (1K)	1 mile
5K	5 miles
10K	
	10 miles
50K	50 miles
100K	
161K	100 miles

The First 100 Certified Aviators
In The Principal Countries Of The World

Compiled from Ch. Faroux and G. Bonnet, *Aéro-Manuel, Répertoire Sportif, Technique et Commercial* (Paris: H. Dunod et E. Pinet, 1914); Fred T. Jane, *All the World's Aircraft* (London: Sampson, Low, Marston & Co., 1913, 1914); Henry Woodhouse, ed., *The Aero Blue Book and Directory of Aeronautic Organizations* (New York: Century Co., 1919); *Direttorio, Pionieri dell' Aeronautica*, and *Aero and Hydro* (20 January and 5 October 1912 and 19 April 1913).

FRANCE

No.	Name	Date of Birth	Machine	Date of License
1	Louis Blériot	1872	Blériot	7 January 1909
2	Glenn Curtiss	1878	Curtiss	7 October 1910
3	Léon Delagrange	1873	Voisin	7 January 1909
4	Robert Esnault-Pelterie	1881	R.E.P	7 January 1909
5	Henry Farman	1874	H. Farman	7 January 1909
5 bis	Capt. Ferdinand Ferber	1862	Voisin	7 January 1909
6	Maurice Farman	1877	M. Farman	18 November 1909
7	Jean Gobron	1885	Voisin	7 October 1909
8	Comte Charles de Lambert	1865	Wright	7 October 1909
9	Hubert Latham	1883	Antoinette	17 August 1909
10	Louis Paulhan	1884	Voisin	17 August 1909
10 bis	Paul Tissandier	1881	Wright	16 September 1909
11	Henry Rougier	1876	Voisin	18 November 1909
12	Alberto Santos-Dumont	1873	Santos-Dumont	7 January 1909
13 *see* 10 bis				
14	Orville Wright	1871	Wright	7 January 1909
15	Wilbur Wright	1867	Wright	7 January 1910
16	Etienne Bunau-Varilla	1890	Voisin	4 November 1910
17	Alfred Leblanc	1869	Blériot	16 December 1909
18	Julien Mamet	1877	Blériot	6 January 1910
19	René Métrot	—	Voisin	6 January 1910
20	Prince G. Bibesco	1880	Blériot	6 January 1910
21	Emile Aubrun	1881	Blériot	6 January 1910

No.	Name	Date of Birth	Machine	Date of License
22	Jacques Balsan	1867	Blériot	6 January 1910
23	Charles S. Rolls	1877	Wright	6 January 1910
24	A. Mortimer Singer	1863	H. Farman	6 January 1910
25	Léon Molon	1881	Blériot	6 January 1910
26	Henri Brégi	1888	Voisin	21 December 1909
27	Jacques de Lesseps	1883	Blériot	6 January 1910
28	Ernest Zens	1878	Blériot	15 January 1910
29	Roger Sommer	1877	Sommer	15 January 1910
30	Claude Grahame-White	1879	Blériot	4 January 1910
31	Michel Efimoff	1881	H. Farman	10 February 1910
32	Georges Chavez	1887	H. Farman	15 February 1910
33	Lt. Felix Camerman	1884	M. Farman	8 March 1910
34	Frederick van Riemsdyck	1887	Curtiss	10 March 1910
35	Edmond Morelle	1880	H. Farman	8 March 1910
36	Baroness Raymonde de Laroche	1884	Voisin	8 March 1910
37	Charles van den Born	1873	H. Farman	8 March 1910
38	Hubert Leblon	1872	Blériot	8 March 1910
39	René Gasnier	1874	Wright	8 March 1910
40	John Moore-Brabazon	1884	Wright	8 March 1910
41	Maurice Herbster	1870	H. Farman	8 March 1910
42	Fernand Delétang	1882	Blériot	5 April 1910
43	André Crochon	1888	H. Farman	5 April 1910
44	Capt. Méderic Burgeat	1864	Antoinette	5 April 1910
45	Lt. Georges Bellenger	1878	Blériot	28 July 1911
46	G. P. Küller	1881	Antoinette	5 April 1910
47	Emile Dubonnet	1883	Tellier	5 April 1910
48	Alfred Frey	1886	H. Farman	5 April 1910
49	Marcel Baratoux	1884	Wright	10 April 1910
50	Prince Nicolas Popoff	1878	Wright	10 April 1910
51	Vincenz Wiesenbach	1880	Wright	10 April 1910
52	Louis Bréguet	1880	Bréguet	10 April 1910
53	Charles Louis Wachter	1874	Antoinette	19 April 1910
54	Léon Morane	1885	Blériot	19 April 1910
55	Georges Legagneux	1882	Sommer	19 April 1910
56	René Toussin	1882	Blériot	19 April 1910
57	Elie A. Mollien	1876	Blériot	19 April 1910
58	Walter de Mumm	1887	Antoinette	19 April 1910
59	Louis Gaubert	1879	Wright	8 May 1910
60	Victor Rigal	1879	Voisin	2 May 1910

No.	Name	Date of Birth	Machine	Date of License
61	Henry Jullerot	1879	H. Farman	2 May 1910
62	Léon Cheuret	1874	H. Farman	2 May 1910
63	Lt. Albert Féquant	1863	H. Farman	2 May 1910
64	René Barrier	1884	Blériot	2 May 1910
65	Capt. Marie Sido	1874	H. Farman	2 May 1910
66	Henri Sallenave	1882	Blériot	2 May 1910
67	Emile E. Bruneau de Laborie	1871	H. Farman	2 May 1910
68	Lt. Paul Aquaviva	1883	Blériot	2 May 1910
69	Alfred de Montigny	1883	Blériot	2 May 1910
70	Hayden Sands	1876	Antoinette	2 May 1910
71	Bertram Dickson	1873	H. Farman	19 April 1910
72	William McArdle	1875	Blériot	19 April 1910
73	Henri Weiss	1880	Blériot	2 May 1910
74	Carl de Cederstrom	1867	Blériot	2 May 1910
75	Douglas Graham-Gilmour	1885	Blériot	19 April 1910
76	Robert Mignot	1882	Voisin	17 May 1910
77	A. Didier	1869	H. Farman	17 May 1910
78	Robert Martinet	1885	H. Farman	17 May 1910
79	Maurice Tetard	1879	H. Farman	17 May 1910
80	Capt. Marie	1870	Blériot	17 May 1910
81	Emile Ladougne	1881	Goupy	17 May 1910
82	Lancelot Gibbs	1883	H. Farman	10 June 1910
83	Louis Wagner	1882	Hanriot	10 June 1910
84	André Taurin	1874	Blériot	10 June 1910
85	Maurice Colliex	1881	Voisin	10 June 1910
86	René Labouchère	1890	Antoinette	10 June 1910
87	Jean Bielovucic	1889	H. Farman	16 June 1910
88	Henri Pequet	1888	Voisin	10 June 1910
89	Capt. Albert Etévé	1880	Wright	10 June 1910
90	Capt. Marconnet	1859	H. Farman	10 June 1910
91	Ernest Paul	1871	Voisin	10 June 1910
92	Louis Gibert	1885	Blériot	10 June 1910
93	André Frey	1886	Sommer	10 June 1910
94	Florentin Champel	1881	Voisin	10 June 1910
95	Marcel Hanriot	1894	Hanriot	10 June 1910
96	Jean Dufour	1889	Voisin	10 June 1910
97	Comdr. Georges Clolus	1867	Antoinette	10 June 1910
98	Vladimir Lebedeff	1884	H. Farman	10 June 1910
99	Marcel Paillette	1884	Sommer	10 June 1910
100	Edmond Audemars	1882	Demoiselle	10 June 1910

No.	Name	Machine	Date of License
1	J.T.C. Moore-Brabazon	Short	8 March 1910
2	Charles S. Rolls	Short Wright	8 March 1910
3	A. Rawlinson	Farman	5 April 1910
4	Cecil S. Grace	Short Wright	12 April 1910
5	George B. Cockburn	Farman	26 April 1910
6	Claude Grahame-White	Blériot	26 April 1910
7	Alec Ogilvie	Short Wright	24 May 1910
8	A. Mortimer Singer	Farman	31 May 1910
9	Samuel F. Cody	Cody	7 June 1910
10	Lt. L.D.L. Gibbs, R.F.A.	Farman	7 June 1910
11	Maurice Egerton	Short Wright	14 June 1910
12	James Radley	Blériot	14 June 1910
13	Alan Boyle	Avis	14 June 1910
14	J. Armstrong Drexel	Blériot	21 June 1910
15	G. C. Colmore	Short	21 June 1910
16	G. A. Barnes	Humber	21 June 1910
17	Capt. Geo. Dawes	Humber	26 July 1910
18	Alliott V. Roe	Roe Triplane	26 July 1910
19	A. E. George	George & Jobling	6 September 1910
20	R. Wickham	Sommer	20 September 1910
21	F. K. McClean	Short	20 September 1910
22	E. K. Davies	Hanriot	11 October 1910
23	Maurice Ducrocq	Farman	1 November 1910
24	J. G. Weir	Blériot	8 November 1910
25	Lt. H. E. Watkins	Howard Wright	8 November 1910
26	Maj. J.D.B. Fulton, R.F.A.	Farman	15 November 1910
28	L. F. Macdonald	Bristol	15 November 1910
29	Lt. R. T. Snowden-Smith	Farman	15 November 1910
30	H. Barber	Valkyrie	22 November 1910
31	Thomas Sopwith	Howard Wright	22 November 1910
32	J. J. Hammond	Bristol	22 November 1910
33	Sydney E. Smith	Bristol	22 November 1910
34	Archibald R. Low	Bristol	22 November 1910
35	R. C. Fenwick	Planes, Ltd.	29 November 1910
36	Capt. A. G. Board	Blériot	29 November 1910
37	Capt. H. F. Wood	Bristol	29 November 1910
38	G. C. Paterson	Paterson	6 December 1910
39	B. G. Bouwens	Blériot	31 December 1910
40	Lt. G. B. Hynes, R.G.A.	Blériot	31 December 1910
41	St. Croix Johnstone	Blériot	31 December 1910
42	Lt. Col. H. R. Cook, R.G.A.	Blériot	31 December 1910
43	Lt. G. H. Barrington-Kennett	Blériot	31 December 1910
44	G.P.L. Jezzi	Jezzi	31 December 1910

No.	Name	Machine	Date of License
45	Lt. R. A. Cammell, R.E.	Bristol	31 December 1910
46	Oscar C. Morison	Blériot	17 January 1911
47	James Valentine	Macfie	17 January 1911
48	H.J.D. Astley	Sommer	24 January 1911
49	Robert Macfie	Macfie	24 January 1911
50	C. Howard Pixton	Roe Triplane	24 January 1911
51	Herbert John Thomas	Bristol	24 January 1911
52	E. Y. Sassoon	Sommer	24 January 1911
53	Geoffrey de Havilland	De Havilland	7 February 1911
54	Capt. D. G. Conner	Farman	7 February 1911
55	J. V. Martin	Farman	7 February 1911
56	A. H. Aitken	Blériot	14 February 1911
57	C.L.A. Hubert	Farman	14 February 1911
58	G. H. Challenger	Bristol	14 February 1911
59	G.R.S. Darrock	Blériot	14 February 1911
60	Archibald Knight	Bristol	14 February 1911
61	C. P. Pizey	Bristol	14 February 1911
62	Louis Maron	Bristol	14 February 1911
63	W. H. Ewen	Blériot	14 February 1911
64	Gustav Hamel	Blériot	14 February 1911
65	Quinto Poggioli	Blériot	28 February 1911
66	Lewis W. F. Turner	Farman	4 April 1911
67	W. Ridley Prentice	Farman	25 April 1911
68	E. C. Gordon England	Bristol	25 April 1911
69	Henry R. Fleming	Bristol	25 April 1911
70	C. C. Turner	Bristol	25 April 1911
71	Comdr. C. R. Samson, R.N.	Short	25 April 1911
72	Lt. A. M. Longmore, R.N.	Short	25 April 1911
73	Lt. W. Parke, R.N.	Bristol	25 April 1911
74	F. Conway Reginald Gregory, R.N.	Short	2 May 1911
75	Lt. Reginald Gregory, R.N.	Short	2 May 1911
76	Maj. Eugene L. Gerrard, R.M.L.I.	Short	2 May 1911
77	E.V.B. Fisher	Hanriot	2 May 1911
78	Hubert Oxley	Hanriot	9 May 1911
79	Harold Blackburn	Bristol	9 May 1911
80	R. C. Kemp	Roe	9 May 1911
81	R. W. Philpott	Bristol	9 May 1911
82	W. H. Dolphin	Hanriot	9 May 1911
83	Lt. C. H. Marks	Farman	9 May 1911
84	Capt. S. D. Massy	Bristol	9 May 1911
85	F. P. Raynham	Roe	9 May 1911
86	J. L. Travers, Jr.	Farman	16 May 1911

No.	Name	Machine	Date of License
87	Edward Hotchkiss	Bristol	16 May 1911
88	Capt. T.C.R. Higgins	Farman	16 May 1911
89	Lt. W. D. Beatty, R.E.	Roe	30 May 1911
90	Lt. R. B. Davies, R.N.	Farman	30 May 1911
91	B. C. Hucks	Blackburn	30 May 1911
92	Capt. H.R.P. Reynolds, R.E.	Bristol	6 June 1911
93	Lt. T. H. Sebag-Montefiore, R.F.A.	Bristol	13 June 1911
94	H. R. Busteed	Bristol	20 June 1911
95	Lt. Col. F. H. Sykes	Bristol	20 June 1911
96	G. Higginbotham	Bristol Curtiss	27 June 1911
97	H. Stanley Adams	Roe	27 June 1911
98	Lt. J. W. Pepper, R. A.	Bristol	27 June 1911
99	Henri Salmet	Blériot	27 June 1911
100	C. Gordon Bell	Hanriot	4 June 1911

GERMANY

No.	Name	Machine	Date of License
1	August Euler	Euler	1 February 1910
2	Hans Grade	Grade	19 February 1910
3	Capt. Paul Engelhard	Wright	15 March 1910
4	Ellery von Gorrissen	Euler	21 April 1910
5	Fridolin Keidel	Wright	17 April 1910
6	Emile Jeannin	Jeannin	27 April 1910
7	Adolf Behrend	Schulze	3 May 1910
8	Eugen Wiencziers	Antoinette	7 May 1910
9	Robert Thelen	Wright	11 May 1910
10	Theodor Schauenburg	Wright	22 July 1910
11	Heinz Reimer Krastel	Blériot	22 June 1910
12	Otto E. Lindpainter	Sommer	14 June 1910
13	Eric Thiele	Euler	10 July 1910
14	Gabriel Poulain	Poulain	15 July 1910
15	Erich Lochner	Euler	11 June 1910
16	Ernst Plochmann	Sommer	23 July 1910
17	Lt. Richard von Tiedemann	Sommer	23 July 1910
18	Herman Dorner	Dorner	27 July 1910
19	Felix Laitsch	Voisin	5 August 1910
20	Simon Brunnhuber	Farman	6 August 1911
21	Oscar Heim	Wright	6 August 1910

No.	Name	Machine	Date of License
22	Walter Lissauer	Grade	7 September 1910
23	Lt. Robert von Mossner	Wright	8 September 1910
24	Lt. Heinrich Haas	Wright	— September 1910
25	Oskar Müller	Wright	— September 1910
26	Lieut. H. Wilberg	Wright	15 September 1910
27	Orla Arntzen	Wright	23 September 1910
28	F. Heidenreich	Heidenreich	23 September 1910
29	Wilhelm Grade	Grade	28 September 1910
30	Bruno Jablonski	Wright	28 September 1910
31	Oswald Kahnt	Grade	28 September 1910
32	Lt. Willy Mente	Wright	28 September 1910
33	Franz Rode	Grade	28 September 1910
34	Gustav Otto	Farman	4 October 1910
35	Bruno Hanuschke	Hanuschke	8 October 1910
36	Friedrich Treitschke	Grade	8 October 1910
37	Heinrich Oelerich	Schutze	21 October 1910
38	Prince Henry of Prussia	—	28 November 1910
39	Julius Artur Weickert	Grade	10 May 1912
40	Bruno Werntgen	Dorner	13 December 1910
41	Wilhelm Hoff	Wright	13 December 1910
42	Raymund Eyring	Huth	13 December 1910
43	Karl Wildt	Blériot	13 December 1910
44	Karl Müller	Farman	24 December 1910
45	Benno König	Farman	29 December 1910
46	Karl Grulich	Harlan	29 December 1910
47	Lt. F. Ferdinand von Hiddessen	Euler	27 January 1910
48	Dr. Josef Hoos	Hoos	17 January 1911
49	Lt. Hammacher	Euler	17 January 1911
50	Werner Dücker	Euler	17 January 1911
51	Lt. Heinrich von Lichtenfels	Euler	17 January 1911
52	C. W. Witterstätter	Aviatik	3 February 1911
53	Bruno Büchner	Aviatik	3 February 1911
54	Fritz Schlüter	Aviatik	3 February 1911
55	Otto Reichhardt	Euler	3 February 1911
56	Hans Roever	Grade	3 February 1911
57	Max Noelle	Grade	3 February 1911
58	Walter de Mumm	Antoinette	(Aero Club de France)
59	Theodor von Flégier	Aviatik	17 February 1911
60	Theodor Meybaum	Grade	17 February 1911
61	Artemy Katzian	Grade	17 February 1911
62	Albert Rupp	Albatros	17 February 1911
63	Georg Schendel	—	—

No.	Name	Machine	Date of License
64	Lt. Konrad Püschel	Wright	23 February 1911
65	Rudolf Kiepert	Wright	24 February 1911
66	Paul Wertheim	Grade	24 February 1911
67	Lugwig Heinz	Albatros	24 February 1911
68	Hans Steinbeck	Grade	27 February 1911
69	Paul Lange	Etrich	28 February 1911
70	Friedrich Krieg	Grade	28 February 1911
71	Ernest Blattman	Wright	28 February 1911
72	Lt. Walter Mackenthun	Farman	7 March 1911
73	Karl Loew	Albatros	10 March 1911
74	Arthur Grünberg	Albatros	24 March 1911
75	Erich Schmidt	Aviatik	17 March 1911
76	Adolf Rentzel	Aviatik	27 March 1911
77	Karl Casper	Etrich	27 March 1911
78	Lt. Edmund Viemela	Blériot	27 March 1911
79	Helmuth Hirth	Etrich	27 March 1911
80	Lt. Reinhold Jahnow	Harlan	10 April 1911
81	Dr. Oskar Wittenstein	Euler	29 April 1911
82	Charles Laemlinn	Aviatik	29 April 1911
83	Hans Roser	Aviatik	29 April 1911
84	Hans Vollmöller	Etrich	15 May 1911
85	Paul Schwandt	Grade	29 May 1911
86	Siegfried Hoffman	Harlan	20 May 1911
87	Gustav Schultze	Schultze	29 May 1911
88	Anthony Fokker	Fokker	17 June 1911
89	Karl Schall	Grade	7 July 1911
90	Lt. Albert Reiche	Grade	November 1912
91	Lt. Felix Lauterbach	Euler	6 July 1911
92	Lt. Wilhelm Wirth	Euler	6 July 1911
93	Count Leopold von Reichenberg	Euler	6 July 1911
94	Franz Oster	Etrich	31 July 1911
95	Lt. Albert Mudra	Euler	8 August 1911
96	Lt. Richard Hartmann	Wright	8 August 1911
97	Gustav Witte	Wright	22 August 1911
98	Lt. Karl Justi	Albatros	22 August 1911
99	Lt. Vogt	Albatros	22 August 1911
100	Walter Hormel	Albatros	24 August 1911

ITALY
Courtesy Italian Embassy, Washington, D.C.

No.	Name	Machine	Date of License
1	Lt. Mario Calderara	Wright	12 September 1910
2	Bartolomeo Cattaneo	Blériot	22 May 1910
3	Stefano Amerigo	—	2 June 1912
4	Lt. Umberto Savoia	Farman	4 July 1910
5	Capt. Ludovico de Filippi	Farman	4 July 1910
6	Baron Leonino Da Zara	Farman	17 August 1910
7	Ugolino Vivaldi Pasqua	Farman	18 August 1910
8	Federico Stucchi	Blériot	19 August 1910
9	Umberto Cagno	Farman	21 August 1910
10	Ernesto Darioli	Blériot	23 August 1910
11	Carletto Pizzagalli	Voisin	2 September 1910
12	Clemente Ravetto	Voisin	2 September 1910
13	Germano Ruggerone	Farman	2 September 1910
14	Ciro Cirri	Blériot	17 September 1910
15	Lt. Enrico Cammarota Adorno	Farman	5 October 1910
16	Lt. Giuseppe Saglietti	Sommer	5 October 1910
17	Mario Faccioli	Faccioli	15 October 1910
18	Umberto Cannoniere	Blériot	17 October 1910
19	Alfredo Cavalieri	Blériot	31 October 1910
20	Lt. Filippo Gazzera	Blériot	31 October 1911
21	Manlio Ginocchio	Blériot	31 October 1910
22	Archimede Lusetti	Blériot	31 October 1910
23	Nino Cagliani	Hanriot	9 November 1910
24	Mario Cobianchi	Farman	13 November 1910
25	Lt. Giulio Gavotti	Farman	17 November 1910
26	Mario Mogafico	Blériot	25 November 1910
27	Giuseppe Rossi	Farman	29 November 1910
28	Ettore Graziani	Farman	13 December 1910
29	Giuseppe Garassini Garbaroni	Farman	22 December 1910
30	Luigi Antonini	—	27 December 1910
31	Giusepe Cei	Caudron	1 January 1911
32	Pietro Cavaglia	Voisin	15 January 1911
33	Ugo de Rossi del Lion Nero	Blériot	23 January 1911
34	Tommaso Surdi	Farman	15 February 1911
35	Lt. Raoul Lampugnani	Farman	20 February 1911
36	Quinto Poggioli	—	20 February 1911
37	Carlo Carabelli	—	3 March 1911
38	Raimondo Marra	Marra	18 March 1911
39	Balilla Battagli	Farman	25 March 1911

No.	Name	Machine	Date of License
40	Sciopio de Campo	Farman	25 March 1911
41	Carlo Maffeis	Blériot	1 May 1911
42	Romolo Manissero	Blériot	3 May 1911
43	Giovanni Ravelli	—	20 May 1911
44	Lt. Leopoldo de Rada	Farman	30 May 1911
45	Capt. Riccardo Moizo	Farman	30 May 1911
46	Gianni Widmer	Blériot	30 May 1911
47	Lt. Leopoldo Stropbino	Farman	1 June 1911
48	Lionello Bonamici	—	6 June 1911
49	Attilio Gelmetti	Hanriot	7 June 1911
50	Gastone Dalla Noce	Blériot	28 June 1911
51	Giovanni Borgotti	Blériot	30 June 1911
52	Roberti di Castelvero	Blériot	30 June 1911
53	Capt. Carlo Maria Piazza	Blériot	30 June 1911
54	Costantino Quaglia	Savary	8 July 1911
55	Lorenzo Santoni	Deperdussin	14 July 1911
56	Umberto Agostoni	Blériot	16 July 1911
57	Francesco Mosca	Blériot	20 July 1911
58	Athos Pintena	Farman	20 July 1911
59	Giulio Brilli Cattarini	Blériot	27 July 1911
60	Achille dal Mistro	—	28 July 1911
61	Guido Nardini	—	3 August 1911
62	Gherardo Baragiola	Caproni	8 August 1911
63	Andrea Poggi	Farman	8 August 1911
64	Agostino de Agostini	Savary	11 August 1911
65	Ettore Marro	Farman	11 August 1911
66	Costantino Biego di Costa Bissara	Caproni 1	16 August 1911
67	Luigino Falchi	Blériot	16 August 1911
68	Alberto Verona	Blériot	16 August 1911
69	Alessandro Guidoni	Farman	18 August 1911
70	Luigi Zorra	—	21 August 1911
71	Gino Gianfelice	Blériot	22 August 1911
72	Arnaldo Porro	—	1 September 1911
73	Ruggero Franzoni	Caproni 1	6 September 1911
74	Angelo Bigliani	Blériot	29 September 1911
75	Celestino Stella	Blériot	2 October 1911
76	Igino Gilbert de Winckel	Farman	7 October 1911
77	Rinaldo Pasquale	Farman	12 October 1911
78	Alfredo de Antonis	Blériot	15 October 1911
79	Umberto Re	Blériot	18 October 1911
80	Francesco Pulvirenti	Blériot	25 October 1911

No.	Name	Machine	Date of License
81	Luigi Bailo	Farman	27 October 1911
82	Clemento Maggiora	Caproni	27 October 1911
83	Alessandro Raffaelli	Farman	27 October 1911
84	Guilio Palma di Cesnola	Blériot	7 November 1911
85	Francesco Vece	Farman	7 November 1911
86	Anselmo Cesaroni	Blériot	22 December 1911
87	Piero Carlo Bergonzi	Caproni	10 January 1912
88	Gaspare Bolla	Blériot	18 January 1912
89	Carlo Graziani	Blériot	18 January 1912
90	Marengo Marenghi	Blériot	18 January 1912
91	Alberto Novellis di Corazze	Blériot	18 January 1912
92	Aldo Bertoletti	Blériot	26 January 1912
93	Giuseppe Colucci	Caproni	27 January 1912
94	Gustavo Moreno	Caproni	30 January 1912
95	Giovanni Sabelli	—	30 January 1912
96	Carlo Vallet	Blériot	2 February 1912
97	Mario Girotto	—	27 February 1912
98	Luigi Berni	Farman	1 March 1912
99	Piero Mandelli	Nieuport	1 March 1912
100	Ernesto Kerbacher	Blériot	9 March 1912

UNITED STATES OF AMERICA

No	Name	Machine	Date of License
1	Glenn H. Curtiss	Curtiss	8 June 1911
2	Frank P. Lahm	Wright	
3	Louis Paulhan	Farman	
4	Orville Wright	Wright	
5	Wilbur Wright	Wright	
6	Clifford B. Harmon	Farman	21 May 1910
7	Thomas S. Baldwin	Curtiss	
8	J. Armstrong Drexel	Blériot	
9	Todd Shriver	Curtiss	17 September 1910
10	Charles F. Willard	Curtiss	
11	James C. Mars	Curtiss	26 August 1910
12	Charles K. Hamilton	Curtiss	26 August 1910
13	John B. Moisant	Blériot	28 and 30 July 1910
14	Charles T. Weymann	Farman	6 June 1910
15	Arthur Stone	Blériot	27 August and 1 September 1910

No.	Name	Machine	Date of License
16	Harry S. Harkness	Antoinette	17 October 1910
17	Eugene Ely	Curtiss	5 October 1910
18	J.A.D. McCurdy	Curtiss	5 October 1910
19	Walter R. Brookins	Wright	18 October 1910
20	Ralph Johnstone	Wright	18 October 1910
21	Arch Hoxsey	Wright	18 October 1910
22	J. Clifford Turpin	Wright	18 October 1910
23	A. L. Welch	Wright	18 October 1910
24	John J. Frisbie	Curtiss	15 October 1910
25	Phillip O. Parmelee	Wright	
26	Frank T. Coffyn	Wright	24 October 1910
27	Lincoln Beachey	Curtiss	7 May 1911
28	Lt. T. G. Ellyson, U.S.N.	Curtiss hydro	2 June 1911
29	Lt. H. H. Arnold, U.S.A.	Wright	6 July 1911
30	Lt. T. de Witt Milling	Wright	6 July 1911
31	Howard W. Gill	Wright	12 July 1911
32	Edson F. Gallaudet	Wright	15 July 1911
33	Harry N. Atwood	Wright	3 July 1911
34	Lee Hammond	Curtiss	24 July 1911
35	W. Redmond Cross	Wright	27 July 1911
36	William Badger	Baldwin	30 July 1911
37	Harriet Quimby	Moisant	1 August 1911
38	Ferdinand E. DeMurias	Moisant	1 August 1911
39	Capt. Paul W. Beck, U.S.A.	Curtiss	3 August 1911
40	William C. Beers	Wright	4 August 1911
41	George W. Beatty	Wright	4 August 1911
42	Hugh Robinson	Curtiss	25 June 1911
43	Cromwell Dixon	Curtiss	6 August 1911
44	Matilde Eleanor Moisant	Moisant	13 August 1911
45	Lt. R. Carrington Kirtland, U.S.A.	Wright	10 August 1911
46	Oscar Allen Brindley	Wright	3 August 1911
47	Leonard Warren Bonney	Wright	3 August 1911
48	Lt. John Rodgers, U.S.N.	Wright	3 August 1911
49	Calbraith Perry Rodgers	Wright	7 August 1911
50	Andrew Drew	Wright	8 August 1911
51	Louis Mitchell	Wright	8 August 1911
52	James J. Ward	Curtiss	8 August 1911
53	Charles C. Witmer	Curtiss	15 August 1911
54	Shakir S. Jerwan	Moisant	26 August 1911
55	Norman Prince	Burgess	26 August 1911
56	Glenn L. Martin	Curtiss	9 August 1911
57	Paul Peck	Rex-Smith	29 and 30 July 1911

No.	Name	Machine	Date of License
58	Harold H. Brown	Wright	7 September 1911
59	Capt. Charles deF. Chandler	Wright	9 September 1911
60	John D. Cooper	Pine	30 August 1911
61	A. B. Lambert	Wright	15 September 1911
62	Lt. John H. Towers	Curtiss	14 September 1911
63	L. E. Holt	Curtiss type	23 August 1911
64	Jesse Seligman	Moisant	24 September 1911
65	Harold Kantner	Moisant	6 September and 14 October 1911
66	Mortimer F. Bates	Moisant	15 October 1911
67	Capt. Geo. W. McKay	Moisant	15 October 1911
68	Phillips Ward Page	Wright	10 October 1911
69	Clifford L. Webster	Wright	10 October 1911
70	Claude Couturier	Wright	3 August and 14 October 1911
71	Beryl J. Williams	Curtiss type	26 August 1911
72	Fred DeKor	Curtiss	14 October 1911
73	Max T. Lillie	Wright	28 October 1911
74	Dr. H. W. Walden	Walden	22 September 1911
75	Albert Elton	Wright	8 October 1911
76	John R. Worden	Moisant	10 and 14 November 1911
77	Clarence A. de Giers	Moisant	6 December 1911
78	Francisco Alvarez	Moisant	7 December 1911

No.	Name	Machine	Date of License
79	Alfred Belognesi	Moisant	17 November and 5 December 1911
80	Antony Jannus	Benoist	27 December 1911
81	Josef Richter	Schneider	27 December 1911
82	Henry D. W. Reichert	Moisant	27 December 1911
83	Horace F. Kearney	Benoist	3 January 1912
84	Arch Freeman	Wright	10 January 1912
85	F. T. Fish	Wright	10 January 1912
86	Frank Champion	Curtiss	10 January 1912
87	Earl Daugherty	Curtiss	10 January 1912
88	Frank M. Stites	Curtiss	10 January 1912
89	Hillery Beachey	Beachey	17 January 1912
90	Lt. J. W. McClaskey	Curtiss	17 January 1912
91	William Hoff	Curtiss	17 January 1912
92	S. C. Lewis	Curtiss	17 January 1912
93	Charles W. Shoemaker	Curtiss	17 January 1912
94	J. B. McCalley	Curtiss	17 January 1912
95	Weldon B. Cooke	Curtiss	17 January 1912
96	Rutherford Page	Curtiss	28 February 1912
97	Lt. Frank N. Kennedy	Curtiss	21 February 1912
98	W. B. Atwater	Curtiss	21 February 1912
99	Albert Mayo	Curtiss	28 February 1912
100	R. C. St. Henry	Curtiss	28 February 1912

Note: It is unclear why no dates were published for the first few licenses or why Curtiss rather than one of the Wrights was granted license No. 1. Possibly they were awarded on "evidence" of an aviator's proficiency presented at Aero Club of America meetings instead of on actual field tests: for example, at a meeting on 5 March 1910, it was simply moved and seconded that Paulhan be granted license No. 3.

Selected Reading List

My Airships, by Alberto Santos-Dumont. (New York: The Century Co., 1904.)

Le Problème de l'Aviation, by J. Armengaud *le jeune.* (Paris: Librairie Ch. Delagrave, 1908.)

L'Aviation: Ses Débuts—Son Développement; de Crête à Crête, de Ville à Ville, de Continent à Continent, by Ferdinand Ferber. (Paris: Berger-Levrault & Cie., 1908.)

The Boys' Book of Airships, by Harry Delacombe. (New York: Frederick A. Stokes Company, 1909.)

How It Flies, by Richard Ferris. (New York: Thomas Nelson & Sons, 1910.)

The Aviator's Companion, by Dick and Henry Farman. (London: Mills and Boon, Limited, 1910.)

L'Aviazione in Italia, by Carlo Montú. (Rome: Armani e Stein, 1911.)

The Aeroplane: Past, Present, and Future, by Claude Grahame-White and Harry Harper. (Philadelphia: J. B. Lippincott Company, 1911.)

Monoplanes and Biplanes, by Grover Cleveland Loening. (New York: Munn & Co., 1911.)

The Conquest of the Air, by Alphonse Berget. (New York: G. P. Putnam's Sons, 1911.)

All About Airships, by Ralph Simmonds. (London: Methuen & Co., Ltd., and Cassell and Company Ltd.; and New York: George H. Doran Company, 1911.)

Heroes of the Air, by Claude Grahame-White and Harry Harper. (New York: Hodder and Stoughton, 1912.)

The Curtiss Aviation Book, by Glenn H. Curtiss and Augustus Post. (New York: Frederick A. Stokes Company, 1912.)

Mes Trois Grandes Courses, by André Beaumont. (Paris: Librairie Hachette & Cie., 1912.)

Les Hydro-Aéroplanes, by Pierre Rivière. (Paris: Librairie Aéronautique, 1912.)

Aviation, by A. E. Berriman. (London: Methuen & Co., Ltd.; New York: George H. Doran Company, 1913.)

The Boys' Book of Aeroplanes, by T. O'B. Hubbard and C. C. Turner. (New York: Frederick A. Stokes Company, 1913.)

All the World's Aircraft, by Fred T. Jane. (London: Sampson, Low, Marston & Co. Ltd., 1909, 1910, 1911, 1912, 1913.)

Guide de l'Aviateur, by Roland Garros. (Paris: Pierre Lafitte & Cie., 1913.)

Taschenbuch der Luftflotten: 1915, by F. Rasch and W. Hormel. (Munich: J. F. Lehmann, 1914.)

Aircraft Year Book, Manufacturers' Aircraft Association, Inc. (Boston: Small, Maynard and Company, 1921.)

La Naissance de l'Aéroplane, by Gabriel Voisin. (Paris: Imprimerie Maréchal, 1928.)

A Narrative History of Aviation, by John Goldstrom. (New York: The Macmillan Company, 1930.)

The World in the Air, Volume II, by Francis Trevelyan Miller. (New York: G. P. Putnam's Sons, 1930.)

Les Héros de l'Air, by Jacques Mortane. (Paris: Librairie Delagrave, 1930.)

The Wright Brothers: Fathers of Flight, by John R. McMahon. (Boston: Little, Brown, and Company, 1930.)

Flying Dutchman, by Anthony H. G. Fokker. (New York: Henry Holt & Company, Inc., 1931.)

Histoire de l'Aéronautique, by Charles Dollfus and Henri Bouché. (Paris: L'Illustration, 1932.)

Flying, by James E. Fechet. (Baltimore: The Williams & Wilkins Company, 1933.)

Our Wings Grow Faster, by Grover Cleveland Loening. (New York: Doubleday, Doran & Company, Inc., 1935.)

The Story of the Winged-S, by Igor I. Sikorsky. (New York: Dodd, Mead and Company, Inc., 1938.)

The Wright Brothers, by Fred C. Kelly. (New York: Harcourt, Brace and Company, Inc., 1943.)

Man's Fight to Fly, by John P. N. Heinmuller. (New York: Aero Print Co., 1945.)

La Grande Histoire de l'Aviation, by René Chambe. (Paris: Ed. Flammarion, 1948.)

La France de M. Fallières, by Jacques Chastenet. (Paris: A. Fayard, 1949.)

Fifty Years of Flying Progress, by Grover Cleveland Loening. (Philadelphia: *Journal of the Franklin Institute,* 1953.)

L'Aviation des Origines à Nos Jours, by Général Hébrard. (Paris: Robert Laffont, 1954.)

Les Conquérants du Ciel, by Roger Sauvage. (Paris: *Le Livre Artistique,* 1960.)

A Pictorial History of Aviation, by the editors of YEAR. (New York: Year, Incorporated, 1961.)

The Saga of Flight, edited by Neville Duke and Edward Lanchbery. (New York: The John Day Co., Inc., 1961.)

The Heritage of Kitty Hawk, by Walter T. Bonney. (New York: W. W. Norton and Company, Inc., 1962.)

Airplanes of the World, by Douglas Rolfe and Alexis Dawydoff. (New York: Simon and Schuster, Inc., 1962.)

The American Heritage History of Flight. (New York: Simon and Schuster, Inc., 1962.)

Hawkins of the Paris Herald, by Eric Hawkins. (New York: Simon and Schuster, Inc., 1963.)

A History of Flight, by Courtlandt Canby. (New York: Hawthorn Books, Inc., 1963.)

The Wright Brothers, by Charles H. Gibbs-Smith. (London: Science Museum, 1963.)

Bell and Baldwin, by J. H. Parkin. (Toronto: University of Toronto, 1964.)

Første Fly over Nordsjøen, by Tryggve Gran. (Oslo: Ernest G. Mortensens Forlag, 1964.)

Match pour la Manche, by Michel Lhospice. (Paris: Editions Denöel, 1964.)

The National Aeronautical Collections, Tenth Edition. (Washington, D.C.: The Smithsonian Institution, 1965.)

The Early Birds, by Arch Whitehouse. (Garden City, N.Y.: Doubleday and Company, Inc., 1965.)

Aeronautics: Early Flying, up to the Reims Meeting, by C. H. Gibbs-Smith, G. W. B. Lacey, W. J. Tuck, and W. T. O'Dea. (London: Science Museum, 1966.)

Mémoires de Roland Garros, by Jacques Quellennec. (Paris: Librairie Hachette & Cie., 1966.)

A Directory and Nomenclature of the First Aeroplanes: 1809 to 1909, by Charles H. Gibbs-Smith. (London: Science Museum, 1966.)

De Clément Ader à Gagarine, by Louis Castex. (Paris: Librairie Hachette & Cie., 1967.)

INDEX

Index

(For illustrations, see the list on page xiii.)